Lecture Notes in Mathematics

A collection of informal reports and seminars
Edited by A. Dold, Heidelberg and B. Eckmann, Zürich

50

Lawrence Zalcman

Massachusetts Institute of Technology, Cambridge

Analytic Capacity and Rational Approximation

1968

Springer-Verlag · Berlin · Heidelberg · New York

Library of Congress Catalog Card Number 68-19414. Printed in Germany. Title No. 7370

PREFACE

The purpose of these notes is to make available in English a reasonably complete discussion of some recent results in rational approximation theory obtained by Soviet mathematicians. More specifically, we shall be concerned with recent theorems of Vitushkin and Melnikov concerning (qualitative) approximation by rational functions on compact sets in the plane. Accordingly, we shall have nothing to say about problems of best approximation, or of approximation on "large" planar sets, or of approximation in the space of n complex variables $(n > 1)$. Each of these subjects is an active discipline in its own right and deserves its own (separate) treatment. On the other hand, since problems of rational approximation have a "local" character, our theorems are related to questions of approximation on regions on Riemann surfaces; however, we shall not pursue that line of thought to any extent.

Since our principal desire is exposition, we have tried to keep the prerequisites for understanding the material at a minimum. A knowledge of basic function theory and functional analysis plus a willingness to pursue a few references given in the text are all that is required. On occasion, it has been convenient to suppress the details of a proof in the interests of exposition; in each such case the reader will supply the missing steps easily. In general, however, when a point has seemed to me obscure, I have chosen to say more rather than less by way of explication.

Although this is not primarily a research paper, it does contain some new material: a few of the examples and the unacknowledged contents of sections 7 and 8 have not appeared in print previously. There are also, as might be expected in a work of this sort, a (small) number of simplifications of proofs, etc. No attempt, however, has been made to take specific notice of such minor improvements. At times I have followed the papers of Melnikov and Vitushkin quite closely; in other instances, the original material has been reorganized considerably. The reader who consults the original papers will easily identify the sections in question.

These notes are based in part on a lecture given at the Brandeis-Brown-M.I.T. joint function algebra seminar at Brown University and on a series of lectures given at Professor Kenneth Hoffman's function algebra seminar at M.I.T. I would like to thank Professor T. W. Gamelin, who first interested me in giving the series of talks mentioned above; his help, encouragement, and enthusiasm at every step of the way have been invaluable. In particular, he read and criticized a preliminary version of the manuscript and made available some very useful notes of his on material covered in sections 11 and 12. I am also very grateful to Dr. John Garnett for many helpful conversations and for permission to incorporate several unpublished results of his into the text; I am especially indebted to him for his careful reading of the final draft. And thanks are due Jane Jordan, who typed the manuscript. Finally, I owe a very real debt of gratitude to the National Science Foundation, which provided me with fellowship support during the preparation of these notes.

CONTENTS

1. INTRODUCTION

Let X be a compact set in the plane and let $R(X)$ denote the set of all functions on X uniformly approximable by rational functions with poles off X. Let $A(X)$ be the set of all functions continuous on X and analytic on X^o, the interior of X. The study of questions relating to $R(X)$ and $A(X)$ is quite old: one can trace interest in rational approximation back to the days of Appell [5],[6], Weierstrass, and Runge [78]. (For further references see p. 141). In recent years, particularly the last two decades, the development of the subject has been especially active. Russian mathematicians have studied rational approximation from the point of view of the constructive theory of functions of one complex variable. The techniques involved in this approach have been, quite naturally, classical; they consist mainly of the construction of suitable kernels and the ingenious application of the complex form of Green's formula. The resulting proofs, while delicate and sometimes quite involved, provide explicit constructions for the approximating functions. These methods have been used with notable success to settle many interesting and important questions [67],[89], [91].

In the United States, recent interest in qualitative rational approximation sprang up from a different corner.

Functional analysts studying certain Banach algebras of functions (uniform algebras, sup norm algebras) were led to examine $R(X)$ and $A(X)$ as interesting special cases of a more general theory. Their approach has been to try to find abstract or semi-abstract proofs for theorems of rational approximation theory. Such efforts have resulted in some beautiful proofs of known theorems [38],[16],[34]. They have met with less success in proving new results (see, however, [107] and [37]).

Recently, Melnikov obtained a necessary and sufficient condition for a point $x \in X$ to be a peak point of $R(X)$ [63],[64]. Since questions concerning peak points arise from considerations of functional analysis rather then function theory, one can regard Melnikov's work as a synthesis of the two approaches mentioned above; here the abstract theory asks the questions, and the constructive methods provide answers. Slightly earlier, Vitushkin [94] had obtained a necessary and sufficient condition on X for $R(X)$ to coincide with $A(X)$. This remarkable achievement generalizes earlier well-known work of Mergelyan and Vitushkin.

With these two results, the theorems of Melnikov and Vitushkin, the theory of rational approximation on compact planar sets approaches a certain completeness. There remain, to be sure, interesting and challenging problems in the subject (see section 16); but the feeling persists that the battle lines for future attacks on these problems have been drawn and that there is every hope that final victory will be attained in the relatively near future. Unfortunately, however, because of the language barrier and the inaccessibility of complete proofs, the results and techniques of [64] and [94] have attracted less attention than they deserve. It is with the hope of rectifying this situation that these notes have been prepared.

A word or two is in order concerning the organization of the material. Sections 2 through 8 deal with material relating to the work of Melnikov, while sections 9 through 14 contain an exposition of Vitushkin's theorem and its amplifications. The reader interested principally in the approximation problem can read section 3 and then skip to section 9. Section 15 contains some applications of "function algebra" methods to rational approximation, including a proof of the fact that $R(X)$ is local; it depends only on section 2. Some open questions are mentioned in section 16, and in two appendices we have collected, for the reader's convenience, relevant facts concerning logarithmic capacity and the removability of singularities for analytic functions. Finally, there is a substantial bibliography with notes; this section is essentially independent of the text.

2. PEAK POINTS

Let X be a compact set in the plane and let $x \in X$. We shall say that x is a <u>peak point</u> for $R(X)$ if there exists $f \in R(X)$ such that (1) $\|f\|_\infty = 1$; (2) $f(x) = 1$; and (3) $|f(y)| < 1$ for $y \in X \setminus \{x\}$. The concept of peak point need not be restricted to the algebra $R(X)$; it makes sense for any sup norm algebra (or, indeed, any collection of complex-valued functions with a common domain of definition). This generalization, however, will not concern us here.

Now fix $x \in X$ and let $L_x(f) = f(x)$, $f \in R(X)$. L_x is a continuous linear functional on $R(X)$ having norm 1. By the Hahn Banach theorem, L_x has a norm-preserving extension $\tilde{L}_x$ to $C(X)$ which, by the Riesz representation theorem, is given by a finite complex Baire measure μ_x supported on X. We then have

$$\tilde{L}_x(f) = \int_X f \, d\mu_x \qquad f \in C(X) .$$

Since

$$1 = L_x(1) = \int_X 1 \, d\mu_x \le \int_X |d\mu_x| = \|\mu_x\| = \|\tilde{L}_x\| = 1 ,$$

it is clear that μ_x is a positive measure of total mass 1. We shall call such a measure a <u>representing measure</u> for x. More precisely, a representing measure for x on $R(X)$ is a Baire probability measure μ, supported on X, such that

$$f(x) = \int_X f \, d\mu$$

for every $f \in R(X)$.

Clearly, the representing measures for x are in one-to-one correspondence with the norm preserving extensions of L_x to $C(X)$; thus, in general, a given point can have many different representing measures. The question of which points have unique representing measures is intimately bound up with peak point considerations. Indeed, we have

Theorem 2.1. (Bishop [13]). Let $X \subset \mathbb{C}$ be compact, $x \in X$. These are equivalent:

(1) x is a peak point for $R(X)$.

(2) x has a unique representing measure supported on X .

(3) Given any (relative) neighborhood U of x , there exists $f \in R(X)$ such that $\|f\|_\infty \leq 1$, $f(x) > 3/4$, and $|f(y)| < 1/4$ if $y \in X \setminus U$.

Proof. (1) => (2) . Obviously, the point mass δ_x represents x on $R(X)$. We must show it is the only probability measure with this property. Suppose $\mu \neq \delta_x$ also represents x . Let $c = \mu(\{x\})$; then $0 \leq c < 1$. The measure $\nu = (1-c)^{-1}(\mu - c\delta_x)$ is a representing measure for x which satisfies $\nu(\{x\}) = 0$. Let f be a function that peaks at x . Then $f^n \to 0$ pointwise boundedly almost everywhere $d\nu$ as $n \to \infty$. By the dominated convergence theorem

$$1 = f^n(x) = \int_X f^n \, d\nu \to 0 ,$$

a contradiction .

(2) => (3) . Let U be given . Choose $v \in C_R(X)$ such that $v \leq 0$, $v < \log(1/4)$ on $X \setminus U$, and $v(x) > \log(3/4)$. It follows from (2) (and the proof of the Hahn Banach theorem) that $v(x) = \sup \{\mathrm{Re}\, g(x) | \ g \in R(X)$ and $\mathrm{Re}\, g \leq v\}$. We may thus assume that $v = \mathrm{Re}\, g$ for some $g \in R(X)$ such that $\mathrm{Im}\, g(x) = 0$. Then the function $f = e^g$ satisfies the conditions of (3). Moreover, $f \in R(X)$, since f can be written as a uniformly convergent power series in g , and $R(X)$ is uniformly closed.

(3) => (1) . Choose a decreasing sequence of neighborhoods $\{U_j\}$ of x such that $U_1 = X$ and $\bigcap_{j=1}^{\infty} U_j = \{x\}$.

Let $V_1 = U_1$. Take f_1 satisfying the condition of (3) for $U = V_1$ and set $g_1 = [f_1(x)]^{-1} f_1$. For $n \geq 2$, choose $V_n \subset U_n \cap V_{n-1}$ and f_n inductively in such a manner that

(a) $|g_j(y)| < 1 + 2^{-3n}$ for $y \in V_n$, $j < n$

and $\|f_n\|_\infty \leq 1$, $f_n(x) > 3/4$, $|f_n(y)| < 1/4$ for $y \notin V_n$.

Let $g_n = [f_n(x)]^{-1} f_n$. Then

(b) $g_n(x) = 1$

(c) $\|g_n\|_\infty \leq 4/3$

(d) $|g_n(y)| < 1/3$ for $y \notin V_n$

Let $f = \sum_{j=1}^{\infty} 2^{-j} g_j$. By (b), $f(x) = 1$. We must show that if $y \neq x$ $|f(y)| < 1$. Suppose, then, that $y \neq x$, and let n be the largest index for which $y \in V_n$. Then, using (a), (c), and (d), we have

$$|f(y)| \leq \left|\sum_{j=1}^{n-1} 2^{-j} g_j(y)\right| + |2^{-n} g_n(y)| + \left|\sum_{j=n+1}^{\infty} 2^{-j} g_j(y)\right|$$

$$\leq (1-2^{-n+1})(1+2^{-3n}) + 2^{-n}(4/3) + 2^{-n}(1/3)$$

$$< 1$$

as required.

As a consequence of 2.1 we obtain

<u>Corollary</u> 2.2. Let $x \in X$. Suppose there exist numbers α, β $(0 < \alpha < \beta < 1)$ such that for any neighborhood U of x there exists a function $f \in R(X)$ satisfying (1) $\|f\|_\infty \leq 1$;

(2) $f(x) > \beta$; and (3) $|f(y)| < \alpha$ if $y \in X \setminus U$. Then x is a peak point for $R(X)$.

<u>Proof</u>. Suppose not. Then, as in the proof of 2.1, there exists a representing measure μ for x such that $\mu(\{x\}) = 0$. Pick a neighborhood U of x such that $\mu(U) < \beta - \alpha$, and let $f \in R(X)$ satisfy the conditions of the corollary. Then

$$\beta < f(x) = \int_X f \, d\mu \leq \left|\int_U f \, d\mu\right| + \left|\int_{X\setminus U} f \, d\mu\right| < (\beta-\alpha) + \alpha = \beta$$

a contradiction .

A constructive proof of 2.2 is given in [39] .

The next two theorems provide some motivation for our study of the peak points of $R(X)$.

<u>Theorem</u> 2.3. Let P be the set of peak points for $R(X)$. Then P is a G_δ . If $x \in X$, there is a representing measure μ_x for x which satisfies $\mu_x(X \setminus P) = 0$.

<u>Theorem</u> 2.4 (Bishop ([13]). If $R(X) \neq C(X)$ then the set of points of X which are <u>not</u> peak points for $R(X)$ has positive (Lebesgue planar) measure.

Theorem 2.3 deserves some comment. The fact that P is a G_δ was noted by Bishop [13] and follows easily from

2.1 ; indeed, if U_n is the set of $x \in X$ for which there exists an f in $R(X)$ satisfying $\|f\|_\infty \leq 1$, $f(x) > 3/4$, and $|f(y)| < 1/4$ for y such that $|x-y| \geq 1/n$, then U_n is open and $\bigcap_{n=1}^{\infty} U_n = P$. The importance of this observation is that it insures that P is a Baire set, thus allowing us to avoid unpleasant problems of measurability. The second part of 2.3 is also due to Bishop, who proved the result for an arbitrary sup norm algebra on a compact metric space. A discussion of this theorem and its generalizations is given in [73]. What we should like to point out is this. By the maximum modulus principle, each function in $R(X)$ attains its maximum on ∂X , the boundary of X . It follows that the restriction map $f \to f|_{\partial X}$ embeds $R(X)$ isometrically in $C(\partial X)$. The reasoning preceding 2.1 now shows that for any $x \in X$ there exists a measure μ_x representing x which is supported <u>on</u> <u>∂X</u> . Since $P \subset \partial X$, 2.3 provides a (strong) generalization of this fact.

We shall postpone the discussion of 2.4 until section 15, where we shall give a proof.

Since $R(X)$ is a Banach algebra it is natural to ask for the identification of its maximal ideal space and Shilov boundary. These questions can be answered without difficulty. It is easy to see that if a point $x \in X$ is linearly accessible from the complement of X then it is a peak point for $R(X)$. Since such points are clearly dense on ∂X , it follows that

∂X is the Shilov boundary of $R(X)$. Moreover, the maximal ideal space of $R(X)$ is X . This follows from the obvious fact that any (nonzero) complex homomorphism of $R(X)$ is determined by its action on the function z .

Finally, let us note explicitly that all the definitions, remarks, and theorems of this section remain valid if we replace $R(X)$ by $A(X)$; the proofs of 2.1, 2.2, and 2.3 remain unchanged while 2.4 becomes trivial. Since $R(X) \subset A(X)$, ∂X is the Shilov boundary of $A(X)$. It is also true [7] that the maximal ideal space of $A(X)$ is X , but this fact is no longer elementary.

3. ANALYTIC CAPACITY

We shall denote the Riemann sphere by S^2. Let U be an open subset of S^2, $\infty \in U$, and suppose the function f is analytic on U. Then f has an expansion

$$f(z) = a_0 + a_1/z + a_2/z^2 + \ldots$$

at $z = \infty$. By definition, $f'(\infty) = a_1$. Alternatively, $f'(\infty) = \lim_{z\to\infty} z(f(z) - f(\infty))$. One should note that $\lim_{z\to\infty} f'(z) \neq f'(\infty)$, since for $a \neq \infty$ the derivative $f'(a)$ is computed with respect to the coordinate $z-a$, while $f'(\infty)$ is computed using the coordinate $1/z$.

Now let K be a compact set in the (finite) plane, and let $\Omega(K)$ denote the unbounded component of $S^2 \setminus K$. The <u>analytic capacity</u> of K is given by

$$\gamma(K) = \sup|f'(\infty)| ,$$

where the sup is taken over all f satisfying

(1) f is analytic on $\Omega(K)$;

(2) $\|f\|_\infty \leq 1$.

We can also require

(3) $f(\infty) = 0$,

since if $g = [f - f(\infty)]/[1 - \overline{f(\infty)}f]$ then g satisfies (1), (2), and (3); and $|g'(\infty)| \geq |f'(\infty)|$. A function satisfying properties (1) - (3) will be called an <u>admissible function</u> for K or, when the context is clear, simply an admissible function. A normal families argument establishes the existence of an admissible function φ such that $\varphi'(\infty) = \gamma(K)$; this function is, in fact, unique [49]. We call φ the <u>Ahlfors function</u> for K. For an arbitrary set S define

$$\gamma(S) = \sup_{K \subset S} \gamma(K) \qquad K \text{ compact .}$$

Analytic capacity has been studied by various authors in some detail ([4],[49],[74],[88]). We shall content ourselves with a discussion of those properties which are more-or-less relevant to the discussion following.

<u>Proposition</u> 3.1. If $S_1 \subset S_2$ then $\gamma(S_1) \leq \gamma(S_2)$. (Monotonicity).

<u>Proposition</u> 3.2. If K_1 and K_2 are compact sets such that $\Omega(K_1) = \Omega(K_2)$ then $\gamma(K_1) = \gamma(K_2)$. In other words, $\gamma(K)$ depends only on the "outer boundary" of K.

<u>Proposition</u> 3.3. Let $a \in \mathbb{C}$. Then $\gamma(aS) = |a|\gamma(S)$, (Homogeneity).

Proposition 3.4. Let $a \in \mathbb{C}$. Then $\gamma(S+a) = \gamma(S)$. (Translation invariance).

Propositions 3.1 and 3.2 are trivial; we leave the easy proofs of 3.3 and 3.4 to the reader.

Proposition 3.5. For any S, $\mathrm{cap}(S) \geq \gamma(S)$. Here $\mathrm{cap}(S)$ denotes the logarithmic capacity of S .

Proof. It is enough to prove the inequality for compact sets. For such sets we have (see Appendix I)

$$\mathrm{cap}(K) = \sup|f'(\infty)| ,$$

where the sup is taken over all functions analytic but not necessarily single-valued on $\Omega(K)$ whose moduli are single-valued and which satisfy $|f(z)| \leq 1$, $f(\infty) = 0$. (The single valuedness of $|f|$ insures that $|f'(\infty)| = \lim_{z\to\infty}|z||f(z)|$ is well defined). It is now clear that $\mathrm{cap}(K) \geq \gamma(K)$.

Proposition 3.6. If C is a continuum, $\mathrm{cap}(C) = \gamma(C) \geq d/4$, where d is the diameter of C .

Proof. Since $\Omega(C)$ is simply connected , $\mathrm{cap}(C) = \gamma(C)$. For continua, it is well-known ([84]) that $\mathrm{cap}(C) \geq d/4$.

Proposition 3.7. Let K be compact and let σ be a rectifiable contour which has winding number 1 around each point of K. Then $\gamma(K) = \sup\left|\frac{1}{2\pi}\int_\sigma f(z)dz\right|$, where the sup is taken over all admissible functions.

Proof. $\frac{1}{2\pi i}\int_\sigma f(z)dz = f'(\infty)$.

Proposition 3.8. Let σ be as above. Then $\gamma(K) \leq \frac{1}{2\pi}$ length(σ). Thus $\gamma(K) \leq \inf \frac{1}{2\pi}$ length (σ), where the inf is taken over the class of all such curves.

Proof. $\gamma(K) = \sup\left|\frac{1}{2\pi}\int_\sigma f(z)dz\right| \leq \frac{1}{2\pi}\int_\sigma |f(z)||dz| \leq \frac{1}{2\pi}\int_\sigma ds =$ $\frac{1}{2\pi}$ length (σ).

Proposition 3.9. Let L be a compact set on a line. Then $\gamma(L) = m(L)/4$, where $m(L)$ = the Lebesgue (linear) measure of L.

Proof. A proof of this result is in [74]. Actually, we will require only the weaker estimate ([4]) $m(L)/4 \leq \gamma(L) \leq m(L)/\pi$. By 3.3 we may assume that L lies on the real axis. It is clear from the preceding proposition that $\gamma(L) \leq m(L)/\pi$. For the other inequality, consider the function

$$g(z) = \frac{e^{f(z)}-1}{e^{f(z)}+1} = \frac{m(L)}{4}\frac{1}{z}+\dots \quad .$$

where

$$f(z) = \frac{1}{2}\int_L \frac{dx}{z-x} = \frac{m(L)}{2z}+ \dots \quad .$$

Now

$$|\operatorname{Im} f(z)| \le \frac{1}{2}\left|\operatorname{Im}\int_L \frac{dx}{(\xi-x)+i\eta}\right| = \frac{1}{2}\int_L \frac{\eta\, dx}{(\xi-x)^2+\eta^2}$$

$$\le \frac{1}{2}\int_{-\infty}^{\infty} \frac{\eta\, dt}{t^2+\eta^2} = \int_0^{\infty} \frac{d\tau}{\tau^2+1} = \arctan \tau \Big|_0^{\infty} = \frac{\pi}{2}$$

Thus $\zeta = f(z)$ maps $S^2\setminus L$ into $|\operatorname{Im}\zeta| \le \frac{\pi}{2}$ so that $e^{f(z)}$ maps $S^2\setminus L$ into the right half plane, and $\|g\|_\infty \le 1$. Since $g(\infty) = 0$, g is admissible.

<u>Proposition</u> 3.10. Let $K_1 \supset K_2 \supset \dots \supset K$ and $\bigcap K_n = K$, where K, K_n are compact sets. Then $\gamma(K_n) \to \gamma(K)$.

<u>Proof</u>. Since $\{\gamma(K_n)\}$ is a decreasing sequence bounded below, its limit obviously exists; moreover, $\lim \gamma(K_n) \ge \gamma(K)$. Let φ_n be the Ahlfors function for K_n. Then $\{\varphi_n\}$ forms

a normal family some subsequence of which converges to a function φ admissible for K. Denote this sequence by $\{\varphi_{n_j}\}$. Let T be a large circle about the origin surrounding K. By 3.7 we have

$$\gamma(K) \geq \left|\frac{1}{2\pi}\int_T \varphi(z)dz\right| = \lim_{n_j\to\infty} \left|\int_T \varphi_{n_j}(z)dz\right| = \lim_{n_j\to\infty} \gamma(K_{n_j}) =$$

$$\lim_{n\to\infty} \gamma(K_n)$$

Thus $\gamma(K) = \lim_{n\to\infty} \gamma(K_n)$. If we use the fact that the Ahlfors function is unique, we can prove even more: namely, that $\varphi_n(z) \to \varphi(z)$ uniformly on compact sets of $\Omega(K)$.

It follows easily from 3.10 that if K is compact, $\gamma(K) = \inf_{U\supset K} \gamma(U)$ where the inf is taken over all open sets containing K. This expresses the fact that, for compact sets, γ is "continuous from above." One might think that, for open sets at least, $\gamma(U) = \gamma(\overline{U})$. Example 3.11 shows that this is not the case.

Example 3.11. Let $L = [0,1]$. Take a sequence of open discs Δ_n of radius r_n such that

(1) $\overline{\Delta}_j \cap \overline{\Delta}_k = \emptyset$ if $j \neq k$,

(2) L is the set of limit points of $\bigcup_{n=0}^{\infty} \Delta_n$,

(3) $\sum r_n < \frac{1}{8}$.

Let $U = \bigcup_{n=0}^{\infty} \Delta_n$; then $\overline{U} = L \cup \bigcup_{n=0}^{\infty} \overline{\Delta}_n$. By 3.1 and 3.9, $\gamma(\overline{U}) \geq \gamma(L) = \frac{1}{4}$. On the other hand, from 3.8 and the definition of analytic capacity for open sets we have

$$\gamma(U) \leq \frac{1}{2\pi} \sum_{n=0}^{\infty} 2\pi r_n = \sum_{n=0}^{\infty} r_n < \frac{1}{8} .$$

Proposition 3.12. (Pommerenke [74]). Let K be a compact set, V(K) the area of K . Then $V(K) \leq \pi[\gamma(K)]^2$.

Proof. Following Pommerenke, we merely sketch the proof. First of all suppose K has finitely many components. Let

$$F(z) = z + \frac{a}{z} + \dots \quad .$$

$$G(z) = z + \frac{b}{z} + \dots \quad .$$

be the (unique) functions functions which map $\Omega(K)$ onto a horizontal slit region and a vertical slit region respectively. Let $\sigma(K) = \frac{a-b}{2}$. In [80], Schiffer showed that $V(K) \leq \pi\sigma(K)$. On the other hand, Ahlfors and Beurling have shown [4] that

$\sqrt{\sigma(K)} \leq \gamma(K)$; thus $V(K) \leq \pi[\gamma(K)]^2$ for compact sets with finitely many components. Applying 3.10, we obtain our result.

There is one more elementary property of analytic capacity we should like to mention, though it will not bear directly on what follows. Define the Painlevé length of a compact set K by

$$\ell(K) = \inf_{\{\Omega_n\}} \lim_{n\to\infty} \text{length}\,(\partial\Omega_n) ,$$

where the inf is taken over all possible sequences $\{\Omega_n\}$ of finitely connected domains (with rectifiable boundaries) which exhaust $\Omega(K)$.

Proposition 3.13. Suppose $\ell(K) < \infty$. Let f be a bounded analytic function on $\Omega(K)$ such that $f(\infty) = 0$. Then

$$f(z) = \int \frac{d\mu(\zeta)}{\zeta - z} ,$$

where μ is a finite (complex) Baire measure supported on ∂K .

Proof. (See [49]). Let $\{\Omega_n\}$ be an exhaustion of $\Omega(K)$ such that $\lim \text{length}(\partial\Omega_n) = \ell(K) + \varepsilon < \infty$. For $z \in \Omega_n$ we have (giving $\partial\Omega_n$ the appropriate orientation)

$$f(z) = (2\pi i)^{-1} \int_{\partial\Omega_n} f(\zeta)(\zeta - z)^{-1}\, d\zeta .$$

Let D be a closed disc whose interior contains K. For large n, $\partial\Omega_n \subset D$. For $g \in C(D)$ define

$$\mu_n(g) = (2\pi i)^{-1} \int_{\partial\Omega_n} f(\zeta) g(\zeta)\, d\zeta .$$

Then

$$\|\mu_n\| = \int_D |d\mu_n| = \int_{\partial\Omega_n} |d\mu_n| = (2\pi)^{-1} \int_{\partial\Omega_n} |f(\zeta)|\, ds \leq$$

$$\frac{\|f\|_\infty \text{ length } (\partial\Omega_n)}{2\pi} .$$

Here we have used the fact that the linear functional μ_n can be identified with the measure $(2\pi i)^{-1} f(\zeta) d\zeta$ restricted to $\partial\Omega_n$ and that their respective norms coincide. Since the $\|\mu_n\|$ are bounded, there exists a cluster point μ of $\{\mu_n\}$ in the weak $*$ topology. It is easy to see that μ is supported on K, in fact on ∂K. Moreover, if $z \in \Omega(K)$ is fixed, the function $h(\zeta) = (\zeta - z)^{-1} \in C(K)$ so that $\int_K (\zeta - z)^{-1} d\mu(\zeta) = \int_D (\zeta - z)^{-1} d\mu(\zeta)$ is a cluster point of the $\int_D (\zeta-z)^{-1} d\mu_n(\zeta)$. Clearly, then,

$$\int_K (\zeta-z)^{-1}\, d\mu(\zeta) = f(z) \ .$$

Finally,

$$\int_K |d\mu| = \|\mu\| \le \varliminf_{n\to\infty} \|\mu_n\| \le (2\pi)^{-1} \|f\|_\infty (\ell(K) + \varepsilon) \ .$$

Using 3.13 we can prove

Proposition 3.14. Suppose $\ell(K) < \infty$. Then

$$\gamma(K) = \sup_\mu \left| \int_K d\mu \right| ,$$

where the sup is taken over all (finite complex Baire) measures μ supported on K which satisfy

$$\sup_{\Omega(K)} \left| \int (\zeta-z)^{-1} d\mu(\zeta) \right| \le 1$$

Proof. Obvious.

An interesting, apparently open, question is whether 3.14 can fail if $\ell(K) = \infty$. Vitushkin has shown [92] that if $\ell(K) = \infty$ then 3.13 is no longer valid.

Finally, we will need the following result due to Mergelyan.

Proposition 3.15. ([81]). Let $K \subset \{|z-z_0| < R\}$. Then for any function

$$f(z) = \frac{a_1}{z-z_0} + \frac{a_2}{(z-z_0)^2} + \dots \quad .$$

admissible for K we have

$$|f(z)| \leq \frac{\gamma(K)}{x-R} \qquad \text{if} \qquad |z-z_0| \geq x > R$$

$$|a_n| \leq e\gamma(K)R^{n-1}n \qquad n = 2,3,\dots \quad .$$

Proof. Fix z_1 such that $|z_1-z_0| \geq x > R$. Let

$$\tilde{f}(z) = \frac{x-R}{z-z_1} \cdot \frac{f(z)-f(z_1)}{1-\overline{f(z_1)}f(z)} = -\frac{x-R}{z-z_0} f(z_1) + \dots \quad .$$

Clearly, $|\tilde{f}(z)| < 1$ if $z \in \Omega(K) \cap \{|z-z_0| < R\}$. Therefore, by the maximum modulus principle, $\tilde{f}$ is an admissible function for K so that $|\tilde{f}'(\infty)| = \lim_{z\to\infty} |z\tilde{f}(z)| = (x-R)|f(z_1)| \leq \gamma(K)$.

This proves the first assertion. To get the coefficient inequality, write

$$a_n = \frac{1}{2\pi i} \int_{|z-z_0|=x>R} f(z)(z-z_0)^{n-1}dz \qquad n = 2,3,\dots \quad .$$

Then $|a_n| \leq \gamma(K)\, x^n (x-R)^{-1}$. The right hand side of this inequality is maximized by setting $x = nR/(n-1)$. Since $(1+1/n)^n < e$, we are done.

It is not hard to improve the coefficient estimate of 3.15; indeed, Vitushkin [88] has shown that $|a_n| \leq A\gamma(K)R^{n-1}$, where A is a universal constant. We shall not need this refinement in the sequel.

4. SOME USEFUL FACTS

This section will be devoted to establishing two rather well-known facts about the Cauchy integral; since references are somewhat scattered, we include them for completeness. It will be convenient to set some notation. Accordingly, let $\Gamma = \{|z| = 1\}$, $\Delta = \{|z| < 1\}$, $\bar{\Delta} = \{|z| \leq 1\}$. This notation will persist throughout the rest of the paper.

<u>Proposition</u> 4.1. Let $F(\zeta) \in C(\Gamma)$, F real . Then for $z \in \Delta$

$$\left|\operatorname{Re} \frac{1}{2\pi i} \int_\Gamma \frac{F(\zeta)}{\zeta - z}\, d\zeta\right| \leq \max_\Gamma |F(\zeta)|$$

<u>Proof</u>. Let $\zeta = e^{i\theta}$. Then

$$\frac{1}{2\pi i} \int_\Gamma \frac{F(\zeta)}{\zeta - z}\, d\zeta = \frac{1}{2\pi} \int_0^{2\pi} F(e^{i\theta}) \frac{e^{i\theta}}{e^{i\theta} - z}\, d\theta$$

Now $\dfrac{e^{i\theta}}{e^{i\theta}-z} = \dfrac{1}{2}\left[\dfrac{e^{i\theta}+z}{e^{i\theta}-z} + 1\right]$ so that $\operatorname{Re}\left[\dfrac{e^{i\theta}}{e^{i\theta}-z}\right] = \dfrac{1}{2}[P_z(\theta) + 1]$

where P_z is the Poisson kernel for z . Thus,

$$\operatorname{Re} \frac{1}{2\pi i} \int_\Gamma \frac{F(\zeta)}{\zeta - z}\, d\zeta = \frac{1}{2}\left[u(z) + \frac{1}{2\pi} \int_0^{2\pi} F(e^{i\theta})\, d\theta\right],$$

where u is the harmonic extension of F to Δ . Since $|u(z)| \leq \max_\Gamma |F(\zeta)|$, the assertion of the proposition is now obvious.

Clearly, if $F \in C(\Gamma)$ is a purely imaginary function, the argument of 4.1 shows that the imaginary part of its Cauchy integral is bounded in modulus by $\max_{\Gamma}|F(\zeta)|$.

<u>Proposition</u> 4.2. Let $F(\zeta) \in C^1(\Gamma)$. Let

$$\hat{F}(z) = \frac{1}{2\pi i}\int_P \frac{F(\zeta)}{\zeta - z}\, d\zeta \ .$$

Then for $|z'| < 1$, $|z''| > 1$ we have

$$\lim_{z',z''\to\zeta_0\in\Gamma}\{\hat{F}(z') - \hat{F}(z'')\} = F(\zeta_0) \ .$$

<u>Proof</u>. By linearity, we can assume F is real. Let $z = \rho e^{i\lambda}$. As before,

$$\frac{1}{i}\cdot\frac{1}{\zeta - z}\, d\zeta = \frac{1}{2}\left[\frac{e^{i\theta}+z}{e^{i\theta}-z} + 1\right] d\theta =$$

$$\frac{1}{2}\left[\frac{1-\rho^2}{1-2\rho\cos(\theta-\lambda)+\rho^2} - i\,\frac{2\rho\sin(\theta-\lambda)}{1-2\rho\cos(\theta-\lambda)+\rho^2} + 1\right] .$$

Let $z' = re^{it}$, $z'' = R\, e^{i\tau}$ $(r < 1 \ , \ R > 1)$.

Then

$$\text{Re }\{\hat{F}(z') - \hat{F}(z'')\} =$$

$$\frac{1}{2}\left\{\frac{1}{2\pi}\int_0^{2\pi} F(e^{i\theta})\left[\frac{1-r^2}{1-2r\cos(\theta-t)+r^2} - \frac{1-R^2}{1-2R\cos(\theta-\tau)+R^2}\right]d\theta\right\}$$

$$= \frac{1}{2}\left\{\frac{1}{2\pi}\int_0^{2\pi} F(e^{i\theta})\left[\frac{1-r^2}{1-2r\cos(\theta-t)+r^2} + \frac{1-\frac{1}{R^2}}{1-\frac{2}{R}\cos(\theta-\tau)+\frac{1}{R^2}}\right]d\theta\right\}$$

$$= \frac{1}{2}\left\{u(re^{it}) + u(\tfrac{1}{R}\, e^{i\tau})\right\},$$

where u is the harmonic extension of F to Δ. Letting $r, R \to 1$ and $t, \tau \to \arg \zeta_0$ we have $\lim\limits_{\substack{z'\to\zeta_0 \\ z''\to\zeta_0}} \text{Re}\{F(z')-F(z'')\} = F(\zeta_0)$.

Similarly,

$$\text{Im}\{\hat{F}(z')-\hat{F}(z'')\} = -\left\{\frac{1}{2\pi}\int_0^{2\pi} F(e^{i\theta})\left[\frac{r\sin(\theta-t)}{1-2r\cos(\theta-\tau)+r^2} - \frac{R\sin(\theta-\tau)}{1-2R\cos(\theta-\tau)+R^2}\right]d\theta\right\}$$

$$= -\left\{\frac{1}{2\pi}\int_0^{2\pi} F(e^{i\theta})\left[\frac{r\sin(\theta-t)}{1-2r\cos(\theta-t)+r^2} - \frac{\frac{1}{R}\sin(\theta-\tau)}{1-\frac{2}{R}\cos(\theta-\tau)+\frac{1}{R^2}}\right]d\theta\right\}$$

$$= \frac{1}{2}\left\{v(re^{it}) - v(\tfrac{1}{R}\, e^{i\tau})\right\},$$

where v is the harmonic conjugate of u such that $v(0) = 0$.

Since F is continuously differentiable, $\lim_{\rho e^{i\lambda} \to \zeta} v(\rho e^{i\lambda})$ exists uniformly for $\zeta \in \Gamma$. Hence $\lim_{z', z'' \to \zeta_0} \operatorname{Im}\{F(z') - F(z'')\} = 0$, which completes the proof.

5. ESTIMATES FOR INTEGRALS

In this section, we shall lay the groundwork for the proof of Melnikov's theorem by proving a sequence of lemmas and theorems, some of which (5.7, 5.8, 5.11) are of considerable interest in themselves. Our treatment follows Melnikov's development [64] quite closely.

Lemma 5.1. Let K be a compact set in the plane. Let f be analytic off K, $f(\infty) = 0$, and $\operatorname{Re} f(z) \leq \alpha$, where $\alpha > 0$. Then $|f'(\infty)| \leq 2\alpha\gamma(K)$.

Proof. Let $f(z) = \sum_{n=1}^{\infty} A_n z^{-n}$ at ∞. Then the admissible function $f_1(z) = f(z)[f(z) - 2\alpha]^{-1}$ has the expansion $f_1(z) = (-A_1/2\alpha)z^{-1} + \ldots$ at ∞. Thus $|f'(\infty)| = 2\alpha|f_1'(\infty)| \leq 2\alpha\gamma(K)$, as required.

Lemma 5.2. Let K be compact, $a \in \Omega(K)$, and let ρ be the distance from a to K. Let f be analytic off K, $\|f\|_\infty \leq 1$. Then $|f'(a)|\rho(\rho-\gamma(K)) \leq 2\gamma(K)$.

Proof. By 3.10, we may assume that K has a boundary consisting of finitely many smooth closed curves. (For if the lemma is proved for this case the general result follows by taking limits). Also, by 3.2, we may assume that each component of K is simply connected. For large R we have, by the Cauchy integral formula,

$$f'(a) = \frac{1}{2\pi i}\int_{|\zeta|=R}\frac{f(\zeta)}{(\zeta-a)^2}\,d\zeta - \frac{1}{2\pi i}\int_{\partial K}\frac{f(\zeta)}{(\zeta-a)^2}\,d\zeta .$$

The first integral on the right is $O(R^{-1})$, so letting $R \to \infty$ we obtain

$$f'(a) = -\frac{1}{2\pi i}\int_{\partial K}\frac{f(\zeta)}{(\zeta-a)^2}\,d\zeta$$

Thus,

$$|f'(a)| = \left|\frac{1}{2\pi i}\int_{\partial K}\frac{f(\zeta)}{(\zeta-a)^2}\,d\zeta\right| = \left|\frac{1}{2\pi i}\int_{\partial K}\left[\frac{f(\zeta)-f(a)}{(\zeta-a)^2} - \frac{f'(a)}{\zeta-a}\right]d\zeta\right| .$$

Now the function

$$F(z) = \frac{f(z)-f(a)}{(z-a)^2} - \frac{f'(a)}{z-a}$$

is obviously analytic off K except perhaps at $z = a$; but at $z = a$ $F(z) = (1/2)f''(a) + \ldots$ so that F is analytic there also. Clearly, $F(\infty) = 0$; and $\|F\|_\infty \leq \frac{2}{\rho^2} + \frac{|f'(a)|}{\rho}$ by the maximum modulus theorem. Therefore, by 3.7,

$$|f'(a)| = \left|\frac{1}{2\pi i}\int_{\partial K} F(\zeta)d\zeta\right| \leq \left[\frac{2}{\rho^2} + \frac{|f'(a)|}{\rho}\right] \gamma(K) .$$

This is just a restatement of the assertion of the lemma.

Note that if $f(a) = 0$ we can strengthen the conclusion of 5.2 to read $|f'(a)|\rho(\rho-\gamma(K)) \leq \gamma(K)$.

<u>Lemma</u> 5.3. Let $K \subset \{r \leq |z| \leq 1\}$ be compact. Denote by K^* the image of K under inversion with respect to the unit circle $(z \to 1/\bar{z})$. Then $r^2\gamma(K^*) \leq 2\gamma(K)$.

<u>Proof</u>. We distinguish two cases. If $0 \notin \Omega(K)$, K must encircle the origin. It is then easy to see that $r \leq \gamma(K)$ and $\gamma(K^*) \leq \frac{1}{r}$ (since $K^* \subset \{|z| \leq \frac{1}{r}\}$) . Hence $r^2\gamma(K^*) \leq r \leq 2r \leq 2\gamma(K)$ as required.

Suppose then that $0 \in \Omega(K)$. Let $f(z)$ be the Ahlfors function for K^* , $\varphi(z) = \overline{f(1/\bar{z})}$. Then $\varphi(0) = 0$ and $\varphi'(0) = \gamma(K^*)$. Applying 5.2 (and the remark following it) to φ (with $a = 0$) we obtain $|\varphi'(0)|r(r-\gamma(K)) \leq \gamma(K)$, so $r^2\gamma(K^*) \leq \gamma(K) + r\gamma(K^*)\gamma(K) \leq 2\gamma(K)$, since $\gamma(K^*) \leq \frac{1}{r}$.

We can now prove

<u>Lemma</u> 5.4. Let $K \subset \bar{\Delta}$, $0 \notin K$. Let K^* be as above. Then

$$\gamma(K \cup K^*) \leq c[\gamma(K) + \gamma(K^*)] ,$$

where c is a universal constant.

Proof. If K separates 0 from ∞ the inequality is obvious, since the l.h.s. is then $\gamma(K^*)$. So suppose K does not separate 0 from ∞. Then we may assume that $\Omega(K) = S^2 \setminus K$. Let $\{K_n\}$ be a decreasing sequence of compact sets each of whose interiors contains K, such that for each n ∂K_n consists of finitely many smooth closed curves, and $\cap K_n = K$. Again, we may assume that $\Omega(K_n) = S^2 \setminus K_n$. It is clear that $K_1^* \supset K_2^* \supset \ldots \supset K^*$ and $\cap K_n^* = K^*$. We will show that

$$\gamma(K \cup K^*) \leq c[\gamma(K_n) + \gamma(K_n^*)]$$

for all n, where c depends neither on n nor on K; the lemma will then follow by 3.10. It is enough to prove that

$$|f'(\infty)| \leq c[\gamma(K_n) + \gamma(K_n^*)]$$

for f admissible for $K \cup K^*$.

Now let n be fixed and f as above be given. We can modify f so that it becomes a smooth function on all of S^2 without disturbing its values on $\overline{S^2 \setminus (K_n \cup K_n^*)}$.

Let us assume this modification accomplished. Let $\varphi_+(z) = f(z) + \overline{f(1/\bar{z})}$, $\varphi_-(z) = f(z) - \overline{f(1/\bar{z})}$. Then φ_+ and φ_- are smooth functions analytic off $K_n \cup K_n^*$ and $\|\varphi_\pm\|_\infty \leq 2$. Also, φ_+ is real on Γ and φ_- is imaginary there. Let

$$\varphi_1(z) = \begin{cases} \varphi_+(z) - \dfrac{1}{2\pi i}\displaystyle\int_\Gamma \dfrac{\varphi_+(\zeta)}{\zeta - z}\,d\zeta & |z| < 1 \\[2ex] -\dfrac{1}{2\pi i}\displaystyle\int_\Gamma \dfrac{\varphi_+(\zeta)}{\zeta - z}\,d\zeta & |z| > 1 \end{cases}$$

$$\varphi_2(z) = \begin{cases} \dfrac{1}{2\pi i}\displaystyle\int_\Gamma \dfrac{\varphi_+(\zeta)}{\zeta - z}\,d\zeta & |z| < 1 \\[2ex] \varphi_+(z) + \dfrac{1}{2\pi i}\displaystyle\int_\Gamma \dfrac{\varphi_+(\zeta)}{\zeta - z}\,d\zeta & |z| > 1 \end{cases}$$

By 4.2 , φ_1 and φ_2 are continuous functions on S^2 . Clearly, φ_1 is analytic off K_n and φ_2 is analytic off K_n^* . By 4.1 we have

$$|\mathrm{Re}\ \varphi_1(z)| \leq 2\|\varphi_+\|_\infty \leq 4$$

$$|\mathrm{Re}\ \varphi_2(z)| \leq 2\|\varphi_+\|_\infty \leq 4$$

Also, $\varphi_1(\infty) = 0$, $|\varphi_2(\infty)| \leq 2$, $\varphi_1 + \varphi_2 = \varphi_+$. Applying 5.1, we get

$$|\varphi_+'(\infty)| \leq |\varphi_1'(\infty)| + |\varphi_2'(\infty)| \leq 8[\gamma(K_n) + \gamma(K_n^*)] .$$

Proceding in a similar manner, we obtain

$$|\varphi_-'(\infty)| \leq 8[\gamma(K_n) + \gamma(K_n^*)] .$$

Since $f = \frac{1}{2}[\varphi_+ + \varphi_-]$, we have

$$|f'(\infty)| \leq 8[\gamma(K_n) + \gamma(K_n^*)] ,$$

as required.

Lemma 5.5. Let $K \subset \{ r \leq |z| \leq 1\}$ be compact. Let f be continuous on $\bar{\Delta}$, analytic on $\Delta \setminus K$, with $\| f\|_\infty \leq 1$. Then

$$\left|\int_\Gamma f(\zeta)\, d\zeta\right| \leq \frac{c}{r^2}\, \gamma(K) ,$$

where c is a universal constant.

Proof. As in 5.4, we may assume that f is smooth on all

of S^2 . Let

$$\varphi_+(z) = \begin{cases} f(z) & |z| < 1 \\ \overline{f(\frac{1}{\bar{z}})} & |z| > 1 \end{cases}$$

$$\varphi_-(z) = \begin{cases} f(z) & |z| < 1 \\ -\overline{f(\frac{1}{\bar{z}})} & |z| > 1 \end{cases}$$

Let

$$\varphi_1(z) = \varphi_+(z) - \frac{1}{2\pi i} \int_\Gamma \frac{f(\zeta)-\overline{f(\zeta)}}{\zeta - z}\, d\zeta$$

$$\varphi_2(z) = \varphi_-(z) - \frac{1}{2\pi i} \int_\Gamma \frac{f(\zeta)+\overline{f(\zeta)}}{\zeta - z}\, d\zeta$$

It is easy to check, using 4.2, that φ_1 and φ_2 are continuous on all of S^2 ; moreover, it is clear that they are analytic off $K \cup K^*$. Applying 4.1, we see that

$$|\text{Im}\ \varphi_1(z)| \leq 3 \qquad |\text{Re}\ \varphi_2(z)| \leq 3 .$$

Therefore, by 5.1,

$$|\varphi_1'(\infty)| \le 6\gamma(K \cup K^*) \qquad |\varphi_2'(\infty)| \le 6\gamma(K \cup K^*) ,$$

so by 5.4

$$|\varphi_1'(\infty)| \le c[\gamma(K)+\gamma(K^*)] \qquad |\varphi_2'(\infty)| \le c[\gamma(K)+\gamma(K^*)] .$$

By 5.3, $\gamma(K^*) \le \frac{2}{r^2}\cdot\gamma(K)$, so that

$$|\varphi_1'(\infty)| \le \frac{c}{r^2}\,\gamma(K) , \quad |\varphi_2'(\infty)| \le \frac{c}{r^2}\,\gamma(K) .$$

Now

$$\varphi_1'(\infty) + \varphi_2'(\infty) = \lim_{z\to\infty} z(\varphi_1(z) + \varphi_2(z))$$

$$= + \frac{1}{\pi i}\int_\Gamma f(\zeta)\,d\zeta .$$

Thus

$$\left|\int_\Gamma f(\zeta)\,d\zeta\right| = \pi|\varphi_1'(\infty) + \varphi_2'(\infty)| \le \frac{c}{r^2}\,\gamma(K) ,$$

as was to be proved.

<u>Exercise</u>. Justify the first sentence in the proof of 5.5.

Lemma 5.6. Let $K \subset \bar{\Delta}$ lie at a positive distance r from Γ. Let f be continuous on $\bar{\Delta}$, analytic on $\Delta \setminus K$, and such that $\|f\|_\infty \leq 1$. Then

$$\left|\int_\Gamma f(\zeta)\, d\zeta\right| \leq \frac{3}{r}\, \gamma(K) .$$

Proof. Let

$$f_1(z) = \frac{1}{2\pi i} \int_\Gamma \frac{f(\zeta)}{\zeta - z}\, d\zeta \qquad |z| < 1$$

$$f_2(z) = -\frac{1}{2\pi i} \int_{|\zeta| = 1-r} \frac{f(\zeta)}{\zeta - z}\, d\zeta \qquad |z| > 1-r .$$

Clearly, f_1 is analytic on Δ and f_2 is analytic on $|z| > 1-r$. By Laurent's theorem, we have

$$f(z) = f_1(z) + f_2(z) \qquad 1-r < |z| < 1 .$$

Thus, we may continue f_2 to a function analytic on $S^2 \setminus K$. It is also clear from the above equality that f_1 can be extended continuously to $\bar{\Delta}$ (and f_2 to $\overline{S^2 \setminus K}$). Now for $z \in \Gamma$ we have, easily, $|f_2(z)| \leq 1/r$. Thus

$$\|f_2\|_{\overline{\Delta}} = \|f-f_1\|_{\overline{\Delta}} \leq \|f\|_{\overline{\Delta}} + \|f_1\|_{\Gamma} = \|f\|_{\overline{\Delta}} + \|f-f_2\|_{\Gamma}$$

$$\leq \|f\|_{\overline{\Delta}} + \|f\|_{\Gamma} + \|f_2\|_{\Gamma} \leq 2 + \frac{1}{r} .$$

Then

$$\left|\int_{\Gamma} f(\zeta)d\zeta\right| = \left|\int_{\Gamma} [f_1(\zeta)+f_2(\zeta)]d\zeta\right| = \left|\int_{\Gamma} f_2(\zeta)d\zeta\right| \leq (2+\tfrac{1}{r})\gamma(K) .$$

Lemma 5.6 can be improved considerably: actually, it remains true when $\frac{3}{r}$ is replaced by an absolute constant, even if we allow K to extend to Γ. This is the content of

<u>Theorem</u> 5.7. Let K be a compact subset of $\overline{\Delta}$. Let $f \in C(\overline{\Delta})$ be analytic on $\Delta \setminus K$ and satisfy $\|f\|_{\infty} \leq 1$. Then

$$\left|\int_{\Gamma} f(\zeta)d\zeta\right| \leq c\gamma(K),$$

where c is a universal constant.

<u>Proof</u>. We may assume that K is bounded by a finite number of smooth closed curves and that $\Omega(K) = S^2 \setminus K$. Moreover, as before, we can assume that f is smooth on all of $\overline{\Delta}$. Let $\beta(z) \in C_R^{\infty}(\overline{\Delta})$ be such that $\beta(z) = 1$ for $z \in \{\frac{2}{3} \leq |z| \leq 1\}$,

$\beta(z) = 0$ for $z \in \{0 \leq |z| \leq \frac{1}{3}\}$, $0 \leq \beta(z) \leq 1$, and $|\frac{\partial\beta}{\partial\bar{z}}| \leq 4$. (Here, as usual, $\frac{\partial}{\partial\bar{z}} = \frac{1}{2}[\frac{\partial}{\partial x} + i\frac{\partial}{\partial y}]$, where $z = x+iy$). Then $\beta f + (1-\beta)f = f$. By the complex form of Green's formula

$$(\beta f)(z) = \frac{1}{2\pi i}\int_{\Gamma\cup\partial K} \frac{(\beta f)(\zeta)}{\zeta - z}\,d\zeta + \frac{1}{2\pi i}\int_{\Delta\setminus K}(\frac{\partial\beta}{\partial\bar{\zeta}})f(\zeta)\frac{1}{\zeta - z}\,d\zeta\wedge d\bar{\zeta}$$

$$= f_1(z) + \varphi_1(z) .$$

Similarly, $(1-\beta)f = f_2 + \varphi_2$. The function φ_1 is continuous on all of S^2, being the convolution of a locally integrable function and a bounded function of compact support ; the same remark holds true for φ_2. We can therefore continue f_j to be continuous on $\bar{\Delta}$ by $f_1 = \beta f - \varphi_1$, $f_2 = (1-\beta)f - \varphi_2$. Let $K_1 = K \cap \{\frac{1}{3} \leq |z| \leq 1\}$, $K_2 = K \cap \{|z| \leq \frac{2}{3}\}$. It is easy to see that f_j is analytic on $\Delta \setminus K_j$. Since $\varphi_1 + \varphi_2 = 0$, $f_1 + f_2 = f$; moreover, $\|f_1\|_\infty \leq \|\beta f\|_\infty + \|\varphi_1\|_\infty \leq c_1$, $\|f_2\|_\infty \leq \|(1-\beta)f\|_\infty + \|\varphi_2\|_\infty \leq c_1$. Applying 5.5 and 5.6 with $r = \frac{1}{3}$ we obtain

$$|\int_\Gamma f(\zeta)d\zeta| \leq |\int_\Gamma f_1(\zeta)d\zeta| + |\int_\Gamma f_2(\zeta)d\zeta|$$

$$\leq 9c_1\gamma(K) + 9c_1\gamma(K) = c\gamma(K) .$$

This completes the proof.

At this point, it is perhaps desirable to offer a few comments on the theorem we have just proved. For one thing, 5.7 is a kind of generalization of Cauchy's theorem. On the other hand, in connection with Morera's theorem, 5.7 shows that (closed) sets of zero analytic capacity are removable for continuous analytic functions (cf. Appendix II). The careful reader will have noticed that we did not actually need the full hypothesis $f \in C(\overline{\Delta})$. In fact, 5.7 remains valid if we require only that f be continuous on and near Γ ; elsewhere, we need only the boundedness of f .

Our next result will be used explicitly in the proof of Melnikov's theorem.

Corollary 5.8. Let $K \subset \{r \leq |z| \leq R\}$ be compact and let f be continuous on the annulus and analytic there off K . Suppose further that $\|f\|_\infty = 1$. Then

$$\left|\int_{|\zeta|=R} f(\zeta)d\zeta - \int_{|\zeta|=r} f(\zeta)d\zeta\right| \leq c(\tfrac{r}{R})\gamma(K) .$$

Here $c(r/R)$ is a constant depending only on r/R ; for t bounded away from 1 , $c(t)$ is bounded away from ∞ .

Proof. By a change of variable we may take $R = 1, r < 1$.

Also, we may once again assume that K has a nice boundary. Then

$$\int_\Gamma f(\zeta)d\zeta - \int_{|\zeta|=r} f(\zeta)d\zeta = \int_{\partial K} f(\zeta)d\zeta$$

by Cauchy's theorem. Let $\beta(z)$ be a real C^∞ function in the annulus such that $\beta(z) = 1$ on Γ $\beta(z) = 0$ on $|z| = r$, $0 \leq \beta(z) \leq 1$, and $\|\frac{\partial\beta}{\partial\bar{z}}\|_\infty \leq c_1(r)$. As in 5.7 we can write $\beta f = f_1 + \varphi_1$, $(1-\beta)f = f_2 + \varphi_2$, where f_1 is analytic on $\Delta \setminus K$, f_2 is analytic on $\{|z| > r\} \setminus K$, $\|f_1\|_\infty \leq c_2(r)$, $\|f_2\|_\infty \leq c_2(r)$, $\varphi_1 + \varphi_2 = 0$. Then 5.8 follows from 5.7 and from the fact that $|\int_{\partial K} g(\zeta)d\zeta| \leq c\gamma(K)$ if $\|g\|_\infty \leq 1$ and g is analytic on $\{|z| > r\} \setminus K$. (The proof of this last fact proceeds exactly like that of 5.7).

<u>Lemma</u> 5.9. Let σ be a circle and let S_1 and S_2 be sets lying (respectively) interior and exterior to σ. Then

$$\gamma(S_1 \cup S_2) \leq c[\gamma(S_1) + \gamma(S_2)],$$

where c is a universal constant.

<u>Proof</u>. We may assume that $S = S_1 \cup S_2$ is a compact set with

boundary consisting of finitely many analytic closed curves. Let φ be the Ahlfors function for S; then φ is continuous on $\Omega(S)$ [33]. Suppose $\sigma = \{|z-z_0| = r\}$. Choose $R \geq 2r$ in such a way that $S_2 \subset \{|z-z_0| < R\}$, and let $\tau = \{|z-z_0| = R\}$. We have

$$\begin{aligned}\gamma(S) = \varphi'(\infty) &= \left|\frac{1}{2\pi i}\int_\tau \varphi(z)dz\right| \leq \\ &\leq \left|\frac{1}{2\pi i}\left[\int_\tau \varphi(z)dz - \int_\sigma \varphi(z)dz\right]\right| + \left|\frac{1}{2\pi i}\int_\sigma \varphi(z)dz\right| \\ &\leq c(r/R)\gamma(S_2) + c_2\ \gamma(S_1) < c[\gamma(S_1)+\gamma(S_2)]\end{aligned}$$

by 5.7 and 5.8 and the fact that $r/R \leq \frac{1}{2}$.

We should remark that we could have avoided using the nonelementary fact about the boundary behavior of the Ahlfors function for a set with analytic boundary; we leave it to the reader to supply an entirely elementary argument.

It is not hard to generalize 5.7 to a theorem for domains with analytic boundary. Let G be a domain bounded by a simple closed rectifiable Jordan curve σ and let $K \subset \bar{G}$ be compact. Define

$$I(G,K) = \sup\left|\int_\sigma f(z)dz\right| ,$$

where the sup is taken over all functions continuous on $\bar{G}$, analytic on $G \setminus K$ such that $\|f\|_\infty \le 1$. We have

<u>Lemma</u> 5.10. Let G_1 and G_2 be domains as above and let $K_1 \subset G$ be closed. Suppose the function $w(z)$ maps G_1 conformally onto G_2 and $0 < m < w'(z) < M < \infty$ for $z \in G_1$. Let $K_2 = w(K_1)$. Then

$$\frac{1}{M} I(G_2,K_2) \le I(G_1,K_1) \le \frac{1}{m} I(G_2,K_2)$$

<u>Proof</u>. Let $\sigma_j = \partial G_j$. Then

$$\int_{\sigma_2} f(w)dw = \int_{\sigma_1} f(w(z))\, w'(z)dz \ .$$

Choose f_1 such that $\int_{\sigma_1} f_1(z)dz \ge I(G_1,K_1) - \varepsilon$, $\varepsilon > 0$. Let $f_2(z) = mf_1(z)/w_1'(z)$ and $f_2(w) = f_2(z(w))$, where $z(w)$ is the function inverse to $w(z)$. Then

$$mI(G_1,K_1) \le \int_{\sigma_1} \frac{mf_1(z)}{w'(z)}\, w'(z)dz + m\varepsilon = \int_{\sigma_2} f_2(w)dw + m\varepsilon$$

$$\le I(G_2,K_2) + m\varepsilon \to I(G_2,K_2) \quad \text{as} \quad \varepsilon \to 0 \ .$$

The other inequality is proved in a similar manner.

Theorem 5.11. Let G be a domain bounded by a simple closed analytic curve σ and let $K \subset \overline{G}$ be compact. Let $f \in C(\overline{G})$ be analytic on $G \setminus K$ and satisfy $\| f\|_\infty \leq 1$. Then

$$\left|\int_\sigma f(z)dz\right| \leq c(G)\gamma(K) ,$$

where $c(G)$ is a constant depending only on G .

Proof. Let G and K be as in the statement of 5.11. There exists a conformal map $w(z)$ of G into Δ ; since σ is an analytic curve $w(z)$ may be extended to a conformal map of a neighborhood $U(G)$ of $\overline{G}$ onto an open disc $U(\Delta)$ containing $\overline{\Delta}$. Clearly, there exist constants m,M such that $0 < m < |w'(z)| < M < \infty$ for $z \in U(G)$. Let $K_1 = w(K)$. By 5.10,

$$I(G,K) \leq \frac{1}{m}\cdot I(\Delta,K_1)$$

$$I(U(\Delta),K_1) \leq M\cdot I(U(G),K)$$

Invoking 5.7, we obtain $I(G,K) \leq \frac{1}{m} c\gamma(K_1)$. Since K lies at a fixed distance from $\partial U(G)$, we have from 5.6 (applied to $U(G)$ instead of Δ)

$$\gamma(K_1) \leq I(U(\Delta),K_1) \leq M\cdot I(U(G),K) \leq M\cdot c_1(G)\gamma(K) .$$

Hence $I(G,K) \leq c(G)\gamma(K)$.

Corollary 5.12. If the sets S_1 and S_2 are separated by an analytic curve σ , then

$$\gamma(S_1 \cup S_2) \leq c(\sigma)[\gamma(S_1) + \gamma(S_2)],$$

where $c(\sigma)$ is a constant depending only on σ .

Proof. This result is the analogue of 5.9 for analytic curves; the proof follows that of 5.9 , using 5.11 .

It is rather remarkable that results such as 5.7 and 5.11 are valid, i.e., that the constants involved do not depend on the compact set in question. One might be led to conjecture that the constant $c(G)$ in 5.11 can be replaced by a universal constant c (independent of G) . Such is not the case. This is the content of

Example 5.13. (Vitushkin [92]). Let $Q = [0,1] \times [0,1]$ be the unit square. Using horizontal lines divide Q into $4n$ congruent rectangles $R_1, R_2, \ldots, R_{4n}$. Let $T = \bigcup_{k=0}^{n-1} R_{4k+1}$ $S = \bigcup_{k=0}^{n-1} R_{4k+3}$. Let G be a simply connected domain with analytic boundary such that $T \subset G$ and $\overline{G} \cap S = \emptyset$. Let

$m > 0$ and set $\lambda(\zeta) = 1$ on T, $\lambda(\zeta) = -1$ on S. Define

$$f(z) = \int_{S\cup T} \frac{\lambda(\zeta)}{\zeta - z}\, d\xi d\eta \qquad (\xi + i\eta = \zeta)$$

An interchange of order of integration gives

$$\int_{\partial G} f(z)dz = \pi i/2 .$$

It is not hard to see that as $n \to \infty$ $\|f\|_\infty \to 0$ ([92]). This shows that $c(G)$ must actually depend on G.

6. MELNIKOV'S THEOREM

We are now prepared to prove Melnikov's theorem. Before proceeding with the proof, however, it will be convenient to set some notation. Accordingly, let $A_n = \{2^{-n-1} \le |z| \le 2^{-n}\}$ and $\Gamma_n = \{|z| = 2^{-n}\}$. If $S \subset \mathbb{C}$ we shall write $CS = S^2 \setminus S$. For $z \in \mathbb{C}$ we set $\rho(z,S) = \inf \{|z-x| : x \in S\}$, the distance from z to S. Finally, $d(S) = \sup \{|z-\zeta| : z,\zeta \in S\}$ is the diameter of S.

<u>Theorem</u> 6.1 (Melnikov). Let $x \in X$ and let $\gamma_n = \gamma(CX \cap \{2^{-n-1} \le |z-x| \le 2^{-n}\})$. Then x is a peak point for $R(X)$ if and only if

$$\sum_{n=0}^{\infty} 2^n \gamma_n = \infty .$$

<u>Proof</u>. For convenience we may suppose $x = 0$.

First, assume that the series above converges. Fix a positive integer N and let f be a rational function whose poles lie off X. We modify f off a neighborhood of X in such a way that

(1) $$\|f\|_{S^2} \le 2\|f\|_X$$

(2) $$\|f\|_{\Gamma_N} \le 2\|f\|_{\Gamma_N \cap X} .$$

Choose a compact set K with smooth boundary so that $X \subset \operatorname{int} K$ and f is analytic on K. Let σ be the boundary of $CK \cap \{|z| \leq 2^{-N}\}$. Then

$$(3) \quad f(0) = \frac{1}{2\pi i} \int_{\Gamma_N} \frac{f(\zeta)}{\zeta} d\zeta - \frac{1}{2\pi i} \int_{\sigma} \frac{f(\zeta)}{\zeta} d\zeta$$

by the Cauchy integral formula (all integrals being taken in the counterclockwise direction). Also, by Cauchy's theorem,

$$(4) \quad \left|\frac{1}{2\pi i} \int_{\sigma} \frac{f(\zeta)}{\zeta} d\zeta\right| = \left|\sum_{n=N}^{\infty} \left[\frac{1}{2\pi i} \int_{\Gamma_n} \frac{f(\zeta)}{\zeta} d\zeta - \frac{1}{2\pi i} \int_{\Gamma_{n+1}} \frac{f(\zeta)}{\zeta} d\zeta\right]\right|,$$

only finitely many summands on the r.h.s. being nonzero. It follows from 5.8 in conjunction with (1) and (4) that

$$(5) \quad \left|\frac{1}{2\pi i} \int_{\sigma} \frac{f(\zeta)}{\zeta} d\zeta\right| \leq c\| f\|_X \sum_{n=N}^{\infty} 2^n \gamma_n ,$$

where c is a universal constant. Consequently, we have from (2), (3) and (5)

$$(6) \quad |f(0)| \leq 2\|f\|_{\Gamma_N \cap X} + c\|f\|_X \sum_{n=N}^{\infty} 2^n \gamma_n .$$

Choosing N so that $\sum_{n=N}^{\infty} 2^n \gamma_n < 1/2c$, we obtain

$$(7) \quad |f(0)| \leq 2\|f\|_{\Gamma_N \cap X} + (1/2)\|f\|_X .$$

This inequality remains valid for all $f \in R(X)$. Now suppose $f \in R(X)$ peaks at 0; then $f^n \in R(X)$ peaks at 0 for every positive integer n. Choosing n large enough, we obtain a contradiction to (7). Thus 0 is not a peak point for $R(X)$.

For the sufficiency, suppose $\sum 2^n \gamma_n = \infty$. Let

$$(8) \quad C_1 = 1 - e\sum_{K=2}^{\infty} \frac{k}{8^{k-1}} > 0 .$$

We shall verify that the α-β condition of 2.2 holds with $\alpha = C_1/10$, $\beta = C_1/9$. Let J_n be a compact subset of $CX \cap A_n$ for which $\gamma(J_n) \geq \gamma_n/2$. Using a finite number of circles we can chop J_n into pieces $J_n^1, \ldots, J_n^m$ (m independent of n) each of which has diameter less than $(1/8)2^{-n-1}$. It follows from 5.9 that there is a universal constant c such that

$$(9) \quad \gamma(J_n) \leq c\sum_{k=1}^{m} \gamma(J_n^k) \qquad n = 0,1,\ldots .$$

For each n pick a set J_n^k for which $\gamma(J_n^k) \geq \gamma(J_n)/mc$ and call it K_n. Then

$$(10) \quad K_n \subset CX \cap A_n ,$$

(11) $d(K_n) \leq (1/8)\, 2^{-n-1}$, and

(12) $\sum 2^n \gamma(K_n) = \infty$.

Let φ_n be the Ahlfors function for K_n and let $z_n \in K_n$. Then

(13) $$\varphi_n(z) = \sum_{k=1}^{\infty} \frac{a_{nk}}{(z-z_n)^k} ,$$

where $a_{n1} = \gamma(K_n)$. Since the circle of radius $d(K_n)$ centered at z_n covers K_n , it follows from 3.15 that

(14) $|a_{nk}| \leq e\gamma(K_n)\, [d(K_n)]^{k-1}\, k$

for $k = 2,3,\ldots$. Moreover, if z lies exterior to a closed disc containing K_n it follows, again from 3.15, that

(15) $|\varphi_n(z)| \leq \gamma(K_n)/\rho(z,K_n)$.

Now let $\delta > 0$ be fixed. Choose N so that

(16) $2^{-N} < (9/20)\cdot C_1\cdot(\delta-2^{-N})$

and then pick M to satisfy

$$(17)\quad 1 < \sum_{n=N}^{M} 2^n \gamma(K_n) < 3 .$$

(This is possible by (11) and (12)). Let

$$(18)\qquad f(z) = (1/9) \sum_{n=N}^{M} \beta_n \varphi_n(z) \quad ,$$

where the β_n are unimodular constants such that $\beta_n \varphi_n(0) \geq 0$. We claim

$$(*)\quad \|f\|_X \leq 1$$

$$(**)\quad |f(z)| < C_1/10 \quad \text{if} \quad |z| \geq \delta$$

$$(***)\quad |f(0)| > C_1/9$$

To obtain (*) , suppose $z \in X \cap A_j$. Then for $n \neq j, j\pm 1$ we have

$$(19)\quad |\varphi_n(z)| \leq \gamma(K_n)/\rho(z,K_n) \leq 2^{n+1} \gamma(K_n)$$

by (15) and (10) . Combining this inequality with the

trivial estimate $|\varphi_k(z)| \leq 1 \quad k = j, j\pm 1$ and invoking (17), we have

$$(20) \quad |f(z)| \leq (1/9) \sum_{n=N}^{M} |\varphi_n(z)| \leq (1/9)[3+2\sum_{n=N}^{M} 2^n \gamma(K_n)] \leq 1 ,$$

as desired.

Next, suppose $|z| \geq \delta$. Since $2^{-N} < \delta$ by (16), we have, from (10),

$$(21) \quad \delta - 2^{-N} \leq \rho(z, A_N) \leq \rho(z, K_N) .$$

Therefore, by (15), (21), and (16)

$$(22) \quad |f(z)| \leq (1/9) \sum_{n=N}^{M} |\varphi_n(z)| \leq (1/9) \sum_{n=N}^{M} \gamma(K_n)/\rho(z, K_n)$$

$$\leq (1/9) \sum_{n=N}^{M} 2^{-n}/(\delta - 2^{-N}) \leq \frac{2}{9} \frac{2^{-N}}{\delta - 2^{-N}} < \frac{C_1}{10} ,$$

which is (**).

It remains to establish (***). Since $|z_n| \geq 8(1/8)\, 2^{-n-1} \geq 8d(K_n)$, we have from (13), (14), and (8)

$$(23) \quad |\varphi_n(0)| \geq \frac{\gamma(K_n)}{|z_n|} - e\gamma(K_n) \sum_{k=2}^{\infty} \frac{k[d(K_n)]^{k-1}}{|z_n|^k}$$

$$= \frac{\gamma(K_n)}{|z_n|} [1-e \sum_{k=2}^{\infty} \frac{k[d(K_n)]^{k-1}}{|z_n|^{k-1}}]$$

$$\geq \frac{\gamma(K_n)}{|z_n|} [1-e \sum_{k=2}^{\infty} \frac{k}{8^{k-1}}]$$

$$= C_1 \frac{\gamma(K_n)}{|z_n|} \geq C_1 2^n \gamma(K_n) .$$

Hence, by (23) and (17),

$$(24) \quad f(0) = (1/9) \sum_{n=N}^{M} \beta_n \varphi_n(0) \geq (C_1/9) \sum_{n=N}^{M} 2^n \gamma(K_n) > C_1/9 ,$$

which is (***) . This completes the proof.

It is worth noticing that Melnikov's theorem can be given a formulation slightly more general than that presented above. Indeed, let $0 < \lambda < 1$ and set $\gamma_n(\lambda) = \gamma(CX \cap \{\lambda^{n+1} \leq |z-x| \leq \lambda^n\})$. Then the proof of Melnikov's theorem shows that

$$\sum_{n=0}^{\infty} \lambda^{-n} \gamma_n(\lambda) = \infty$$

is also necessary and sufficient for $x \in X$ to be a peak point for $R(X)$.

7. FURTHER RESULTS

In general, it is not easy to verify the condition of 6.1 directly; therefore, it is useful to have available a truly geometric condition for a point to be a peak point for $R(X)$. For $x \in X$ and $r > 0$ let $d(r)$ be the supremum of the diameters of the components of $CX \cap \{|z-x| < r\}$. In [39] , Gonchar proved that if

$$\overline{\lim}\, d(r)/r > 0 \qquad (r \to 0)$$

then x is a peak point for $R(X)$. Using Melnikov's theorem we can prove a more general result due to P. C. Curtis [114].

Theorem 7.1. Let X be compact , $x \in X$. Then if

$$\overline{\lim_{r\to 0}}\ \frac{\gamma(CX \cap \{|z-x| < r\}}{r} > 0 ,$$

x is a peak point for $R(X)$.

Proof. For convenience, take $x = 0$. Let $r_n \downarrow 0$ be such that $\gamma(CX \cap \{|z| < r_n\})/r_n > \varepsilon > 0$ and take $0 < \lambda < \varepsilon/c$, where c is the constant of 5.9. Then for large n we have $\lambda^{k+1} \leq r_n < \lambda^k$, where, of course, k depends upon n. Applying 5.9 twice, we obtain

$$\lambda^{k+1}\varepsilon \leq r_n \varepsilon \leq \gamma(CX \cap \{|z| < r_n\})$$

$$\leq c\,\gamma(CX \cap \{|z| \leq \lambda^{k+2}\}) + c\gamma(CX \cap \{\lambda^{k+2} \leq |z| \leq \lambda^k\})$$

$$\leq c\lambda^{k+2} + c^2\,\gamma_k(\lambda) + c^2\gamma_{k+1}(\lambda) \quad ,$$

where $\gamma_n(\lambda) = \gamma(CX \cap \{\lambda^{n+1} \leq |z| \leq \lambda^n\})$. Hence,

$$\frac{\lambda^{k+1}}{c} \cdot \left(\frac{\varepsilon}{c} - \lambda\right) \leq \gamma_k(\lambda) + \gamma_{k+1}(\lambda) .$$

It follows that

$$\sum_{n=0}^{\infty} \lambda^{-n}\,\gamma_n(\lambda) = \infty .$$

By the remark following the proof of Melnikov's theorem we are done.

It is easy to see that the condition of 7.1 is not necessary for x to be a peak point.

Exercise. Prove this assertion.

We now turn to some corollaries of Melnikov's theorem.

Theorem 7.2. Let $x \in X$ and let ω_n be the area of $CX \cap \{2^{-n-1} \leq |z-x| \leq 2^{-n}\}$. If $\sum 4^n \omega_n = \infty$ then x is a peak point for $R(X)$.

Proof. $\infty = \sum 4^n \omega_n = \sum (2^n\sqrt{\omega_n})^2 \leq \sum (\sqrt{\pi})(2^n\sqrt{\pi}\, \gamma_n) = \pi \sum 2^n \gamma_n$. Now apply Melnikov's theorem. (We have used the fact that $\omega_n \leq \pi 4^{-n}$ and 3.12).

Theorem 7.3. Let $x \in X$ and let $\omega(r)$ be the area of $CX \cap \{|z-x| < r\}$. Suppose

$$\int_0^1 \frac{\omega(r)}{r^3}\, dr = \infty .$$

Then x is a peak point for $R(X)$.

Proof. We have

$$\infty = \int_0^1 \frac{\omega(r)}{r^3}\, dr = \int_0^1 \omega(r)\, d(-\tfrac{1}{2}\, \tfrac{1}{r^2}) = -\tfrac{1}{2}\Big[\frac{\omega(r)}{r^2}\Big|_0^1 + \int_0^1 \frac{d\omega(r)}{r^2}\Big].$$

Clearly, the first term of the r.h.s. is bounded in absolute value by $\pi/2$. On the other hand,

$$\int_0^1 \frac{d\omega(r)}{r^2} = \sum_{n=0}^{\infty} \int_{2^{-n-1}}^{2^{-n}} \frac{d\omega(r)}{r^2} \leq 4 \sum_{n=0}^{\infty} 4^n \int_{2^{-n-1}}^{2^{-n}} d\omega(r) = 4 \sum_{n=0}^{\infty} 4^n \omega_n .$$

Thus, 7.3 follows from 7.2.

Although the (equivalent) conditions 7.2 and 7.3 are sufficient for 0 to be a peak point, they are far from necessary. Indeed, let $\Delta_n = \{|z-3\cdot 2^{-n-2}| < n^{-1}2^{-n}\}$ and set $X = \bar{\Delta} \setminus \bigcup_{n=4}^{\infty} \Delta_n$. Then $CX \cap A_n = \Delta_n$ $n = 4,5,\ldots$ so that, for $x = 0$, $\gamma_n = n^{-1}2^{-n}$. Now $\sum 2^n \gamma_n = \sum n^{-1} = \infty$; but $\omega_n = n^{-2}4^{-n}\pi$ so that $\sum 4^n \omega_n = \pi \sum n^{-2} < \infty$.

One should not fail to mention the relation of the results we have discussed to the classical theory of harmonic functions. Let D be a planar domain and let u be a real continuous function on ∂D. To each such u we can associate, in a canonical manner, a function $\hat{u}$ (continuous and) harmonic on D. A point $x \in \partial D$ is a <u>regular point</u> for the Dirichlet problem on D if $\hat{u}(z) \to u(x)$ as $z \to x$ from within D for every $u \in C_R(\partial X)$. Wiener [105] showed that $x \in \partial D$ is a regular point for the Dirichlet problem on D if and only if

$$\sum_{n=0}^{\infty} \frac{n}{\log(1/c_n)} = \infty ,$$

where $c_n = \operatorname{cap}(CD \cap \{\lambda^{n+1} \leq |z-x| \leq \lambda^n\}$, $0 < \lambda < 1$. This condition is independent of the choice of λ. (Actually, Wiener's condition was phrased differently, since the definition of logarithmic capacity which he used is not the one in common currency today). In particular, any of the (inequivalent) conditions $\sum \lambda^{-n} c_n = \infty$ implies that x is a regular point (see [84]). Moreover, the conditions of 7.2 and 7.3 are sufficient to insure that x be a regular point for D ([84]).

A decade after Wiener, Keldysh [57] observed that one could characterize the regular points for a domain in terms of the peaking properties of an appropriate space of harmonic functions. More specifically, let $\mathcal{D}(\overline{D})$ be the set of all functions continuous on D and harmonic on D. Keldysh showed that $x \in \partial D$ is a regular point for the Dirichlet problem on D if and only if x is a peak point for $\mathcal{D}(\overline{D})$; a simple proof of this fact is in [32]. Since $\mathcal{D}(\overline{D}) \supset \operatorname{Re} R(D)$, it is clear that any condition which implies that $x \in \overline{D}$ is a peak point for $R(\overline{D})$ implies x is a peak point for $\mathcal{D}(\overline{D})$ and hence a regular point for the Dirichlet problem on D. Thus 7.2 and 7.3 strengthen known results from potential theory.

8. APPLICATIONS

Melnikov's theorem gives us useful information even in the simplest (nontrivial) cases. Consider, for example, a compact set X constructed as follows. Delete from $\bar{\Delta}$ a sequence of open discs $\Delta_n = \{|z-x_n| < r_n\}$, where

(1) $1 > x_1 > x_2 > \dots \to 0$;

(2) $x_1 + r_1 < 1$;

(3) $x_{n+1} + r_{n+1} < x_n - r_n$ for all n.

Then $X = \bar{\Delta} \setminus \bigcup_{n=1}^{\infty} \Delta_n$. We shall call such an X a set of type (L). Thus, if X is of type (L), it is the closure of a domain (X^o) obtained by deleting from the open unit disc the origin and a sequence of pairwise disjoint closed discs centered on the positive real axis and clustering at 0. It is clear that $\sum_{n=1}^{\infty} r_n < \infty$. It is also obvious that each point of $\partial X \setminus \{0\}$ is a peak point for $R(X)$. The natural question to ask is: when is 0 a peak point for $R(X)$? Melnikov's theorem allows us to give this question a complete answer.

Theorem 8.1. Let X be of type (L). Then 0 is a peak point for $R(X)$ if and only if $\sum_{n=1}^{\infty} (r_n/x_n) = \infty$.

Proof. Let $I = CX \cap [0,1]$, $I_n = I \cap A_n$, $J_n = I \cap \Delta_n$. First we show that $\int_I x^{-1}\,dx$ diverges with $\sum (r_n/x_n)$. Indeed, we have

$$\sum_{n=1}^{\infty} \frac{r_n}{x_n} \leq \sum_{n=1}^{\infty} \frac{2r_n}{x_n+r_n} \leq \sum_{n=1}^{\infty} \int_{J_n} \frac{dx}{x} = \int_I \frac{dx}{x} .$$

Hence, for the sufficiency it is enough to show that 0 is a peak point if $\int_I x^{-1}\,dx = \infty$. Suppose, then, that the integral diverges. Then

$$\infty = \int_I x^{-1}\,dx = \sum_{n=0}^{\infty} \int_{I_n} x^{-1}\,dx \leq \sum_{n=0}^{\infty} 2^{n+1} \int_{I_n} dx$$

$$= 2 \sum_{n=0}^{\infty} 2^n \text{ length } (I_n) = 8 \sum_{n=0}^{\infty} 2^n \gamma(I_n)$$

$$\leq 8 \sum_{n=0}^{\infty} 2^n \gamma_n ,$$

where we have used 3.1 and 3.9. Thus $\sum 2^n \gamma_n = \infty$, and 0

is a peak point. For the other direction, we do not need to invoke Melnikov's theorem. Suppose $\sum r_n/x_n < \infty$. Then $r_n/x_n \to 0$, so that for some integer K $(0 < K < \infty)$ $r_n/(x_n-r_n) < K(r_n/x_n)$ for all n. Consider the (complex) measure $d\mu(\zeta) = (2\pi i)^{-1}\zeta^{-1}d\zeta$ on ∂X, where the integration is carried out, as usual, counterclockwise on Γ and clockwise on $\partial X \setminus \Gamma$; μ has finite total variation since

$$\int_{\partial X} \frac{ds}{|\zeta|} \leq 2\pi + \sum_{n=1}^{\infty} \frac{2\pi r_n}{x_n - r_n} \leq 2\pi(1 + K\sum_{n=1}^{\infty} \frac{r_n}{x_n}) < \infty .$$

Clearly, $|\mu|(\{0\}) = 0$. Since

$$\int_{\partial X} r(\zeta)\, d\mu(\zeta) = r(0)$$

for every rational function with poles off X, we have

$$\int_{\partial X} f(\zeta)\, d\mu(\zeta) = f(0)$$

for every $f \in R(X)$. Suppose $f \in R(X)$ peaks at 0. Then $f(0) = 1$ and $|f(\zeta)| < 1$, $\zeta \neq 0$. Thus $f^n(\zeta) \to 0$ pointwise boundedly almost everywhere $d|\mu|$. But then, by the dominated convergence theorem,

$$1 = f^n(0) = \int f^n(\zeta)\, d\mu(\zeta) \to 0 \quad \text{as} \quad n \to \infty ,$$

a contradiction. Hence no function in $R(X)$ can peak at 0.

Theorem 8.1 admits immediate generalization. For instance, if X is the closure of a domain whose boundary consists of the origin and a union of disjoint circles accumulating only at 0, and if the number of components of $CX \cap A_n$ is bounded independently of n, then 8.1 holds. (of course we must replace x_n by $|x_n|$). Since the condition for a point to be a peak point is local (i.e., depends only on the structure of X near the point in question) we need only require the above boundedness condition near 0.

According to 8.1, 0 fails to be a peak point for a set X of type (L) if and only if the Cauchy measure for 0, $(2\pi i)^{-1} \zeta^{-1} d\zeta$, exists as a finite measure on ∂X. It is tempting to conclude that this situation persists whenever ∂X is rectifiable. More precisely, suppose X is the closure of a domain whose rectifiable boundary consists of the origin and a union of disjoint circles accumulating only at 0; one might conjecture that if 0 fails to be a peak point for $R(X)$ then Cauchy measure for 0 exists as a finite measure on ∂X. However, this is not the case. In fact, we have the following

Example 8.2. Let $\Delta_n = \{|z-4^{-n}-4^{-2n}| < 4^{-2n}\}$, $n = 1,2,\ldots$. Then $X_0 = \bar{\Delta} - \bigcup_{n=1}^{\infty} \Delta_n$ is a set of type (L) and $\sum (r_n/x_n) = \sum 4^{-2n}/(4^{-n}-4^{-2n}) < \infty$. Hence 0 is not a peak point for $R(X_0)$. It is clear, therefore, that if Y is compact and $Y \supset X_0$ then 0 is not a peak point for $R(Y)$. For each n let $D_{n,j}$ $(j = 1,2,\ldots,k_n)$ be a collection of open discs with radii $r_{n,j}$ such that

(1) $\bar{D}_{n,j} \subset \Delta_n$ for all n,j

(2) $\bar{D}_{n,j} \cap \bar{D}_{n,k} = \emptyset$ unless $j = k$

(3) $n^{-2} < \sum_{j=1}^{k_n} r_{n,j} < 2n^{-2}$ for each n

Let $Y = \bar{\Delta} \setminus \bigcup_{n,j} D_{n,j}$. Then $Y = \overline{Y^o}$, ∂Y consists of the origin and a union of disjoint circles accumulating only at 0, and ∂Y is rectifiable since

$$\sum_{n=1}^{\infty} \sum_{j=1}^{k_n} r_{n,j} \leq 2 \sum_{n=1}^{\infty} \frac{1}{n^2} = \frac{\pi^2}{3} .$$

Since $Y \supset X_0$, 0 is not a peak point for $R(Y)$. On the other hand,

$$\sum_{n=1}^{\infty} \sum_{j=1}^{k_n} \frac{r_{n,j}}{|x_{n,j}|} \geq \sum_{n=1}^{\infty} 4^n \sum_{j=1}^{k_n} r_{n,j} > \sum_{n=1}^{\infty} \frac{4^n}{n^2} = \infty$$

so that the measure $(2\pi i)^{-1} \zeta^{-1} d\zeta$ does not have finite total variation.

At this point, it is perhaps worthwhile to note that the notion of analytic capacity is quite crucial for Melnikov's theorem: even in the simplest cases it is not possible to replace analytic capacity by the more familiar logarithmic capacity. This is shown by

Example 8.3. Let Y be the compact set obtained by deleting from $\overline{\Delta}$ the open discs Δ_n where

$$\Delta_{2k+1} = \{ |z - \frac{1}{2^{k+2}} (\frac{2k^2+8}{k^2+3}) | < \frac{1}{2^{k+2}(k^2+3)} \}$$

$$\Delta_{2(k+1)} = \{ |z - \frac{1}{2^{k+2}} (\frac{4k^2+10}{k^2+3})| < \frac{1}{2^{k+2}(k^2+3)} \} .$$

$k = 0,1,2,\ldots$. Then $\overline{\Delta}_{2k+1} \cup \overline{\Delta}_{2(k+1)}$ lies in the interior of $A_k = \{2^{-k-1} \leq |z| \leq 2^{-k}\}$, and Y is a set of type (L). We claim $\sum 2^n \gamma_n < \infty$ but $\sum 2^n c_n = \infty$, where c_n denotes the logarithmic capacity of $CX \cap A_n$. From 3.8 it is clear that $\gamma_n \leq 2^{-(n+1)} (n^2+3)^{-1}$, so that $\sum 2^n \gamma_n < \infty$. On the other hand,

$$c_n \geq \operatorname{cap}\left(\left[\frac{1}{2^{n+2}}\left(\frac{2n^2+7}{n^2+3}\right), \frac{1}{2^{n+2}}\left(\frac{2n^2+9}{n^2+3}\right)\right]\right.$$

$$\left.\cup \left[\frac{1}{2^{n+2}}\left(\frac{4n^2+9}{n^2+3}\right), \frac{1}{2^{n+2}}\left(\frac{4n^2+11}{n^2+3}\right)\right]\right)$$

$$= \frac{1}{2^{n+2}(n^2+3)} \operatorname{cap}([2n^2+7, 2n^2+9] \cup [4n^2+9, 4n^2+11])$$

$$= \frac{1}{2^{n+2}(n^2+3)} \operatorname{cap}([-n^2-2, -n^2] \cup [n^2, n^2+2])$$

by the monotonicity, homogeneity, and translation invariance of logarithmic capacity. If we use the fact that the logarithmic capacity of a compact set coincides with its transfinite diameter (see Appendix I), it is not hard to show that

$$\operatorname{cap}([-n^2-2, -n^2] \cup [n^2, n^2+2]) = \sqrt{n^2+1} > n$$

so that

$$\infty = \frac{1}{4} \sum_{n=1}^{\infty} \frac{n}{n^2+3} < \sum_{n=1}^{\infty} 2^n c_n ,$$

as required. By a remark in the preceding section it follows that 0 is a regular point for the Dirichlet problem for

Y^o (in fact, a point of stability, see [57]); yet 0 fails to be a peak point for $R(Y)$. Note, however, that if $CX \cap A_n$ consists of a single component for each n we can replace γ_n by c_n in Melnikov's theorem ; this follows from 3.6.

Finally, let us remark that we have said nothing about the characterization of the peak points of $A(X)$; this more difficult problem remains open (see 16.4). Of course, Melnikov's theorem gives a sufficient condition, and the condition is necessary as well when $R(X) = A(X)$. It is to the question of when these two algebras coincide that we now turn.

9. THE PROBLEM OF RATIONAL APPROXIMATION

Let $X \subset \mathbf{C}$ be compact. What can we say about the functions in $R(X)$? Clearly, they belong to $C(X)$; moreover, they are analytic on X^o , the interior of X . (This follows from the fact that each function in $R(X)$ is a uniform limit on X of functions analytic on X^o). Denote by $A(X)$ the algebra of functions in $C(X)$ which are analytic on X^o ; then $A(X) = C(X)$ if and only if X^o is empty. We have observed that $R(X) \subset A(X)$. Can this inclusion be proper? What if $X^o = \emptyset$? When is $R(X) = A(X)$? These natural questions are basic to the qualitative theory of rational approximation on compact planar sets. We shall treat them in some detail in the succeeding sections.

The best understanding we have of the problems which arise in rational approximation comes from examples (and counterexamples). Accordingly, most of this section will be devoted to a discussion of previously known results, with a special emphasis on examples. Our first theorem is an essentially trivial but enormously useful observation due to Runge.

<u>Theorem</u> 9.1. Let f be analytic in a neighborhood U of X . Then $f|_X \in R(X)$, where $f|_X$ denotes the restriction of f to X .

Proof. Let $\sigma \subset U$ be a rectifiable contour that has winding number 1 about each point of X. Then

$$f(z) = \frac{1}{2\pi i} \int_\sigma \frac{f(\zeta)}{\zeta - z} d\zeta \qquad z \in X .$$

Since the right hand side is a Riemann integral it is (for fixed z) a limit of sums of the form $(2\pi i)^{-1} \sum_{j=1}^{n} c_j (\zeta_j - z)^{-1}$, $\zeta_j \in \sigma$. Since X is contained in the interior of the (not necessarily connected) set bounded by σ, it is easy to see that the Riemann sums converge uniformly on all of X.

Let $\mathcal{O}(X)$ denote the algebra of all functions analytic on (a neighborhood of) X, and let $\overline{\mathcal{O}(X)}$ be the closure of $\mathcal{O}(X)$ with respect to the uniform norm on X. Of course, $R(X) \subset \overline{\mathcal{O}(X)}$. The content of Theorem 9.1 is that $\mathcal{O}(X) \subset R(X)$ so that $R(X) = \overline{\mathcal{O}(X)}$. This situation does not persist in $\mathbb{C}^n$ $(n > 1)$.

A second useful observation is this. Let T be a component of $S^2 \setminus X$ and let $\alpha, \beta \in T$, $\alpha \neq \infty$. Then $(z-\alpha)^{-1}$ can be approximated on X uniformly by polynomials in $(z-\beta)^{-1}$ (in z, if $\beta = \infty$). This is clear for α near β and follows in general from a standard connectedness argument. Thus, in studying rational approximation we may fix a point p_n in each component T_n of $S^2 \setminus X$ and then consider only rational functions whose poles lie in the set $\{p_n\}$. In particular, if $T_0 = \Omega(X)$ we may choose $p_0 = \infty$. Suppose now that $S^2 \setminus X = \Omega(X)$. Then every function in $R(X)$ can be approximated uniformly by rational functions whose only pole lies at ∞, i.e., by polynomials. Thus, for sets which do not divide the plane, polynomial approximation and rational approximation are (qualitatively) the same.

We begin with a brief discussion of the situation in which X is "simply connected," i.e., for which $S^2 \setminus X$ is connected. We shall delay the proofs of the results stated below, since these will appear as easy consequences of a later theorem (see section 13).

Example 9.2. Let $X = [a,b]$, an interval on the real axis. The Weierstrass approximation theorem implies that $R(X) = C(X)$ in this case. Similarly, if X is any compact set on the line $R(X) = C(X)$, since if $X \subset [a,b]$ any function in $C(X)$ can be extended to a function in $C([a,b])$.

Example 9.3. Recall that Δ denotes the open unit disc $\{|z| < 1\}$ and $\overline{\Delta}$ is its closure, $\{|z| \leq 1\}$. Let $X = \overline{\Delta}$. Then $A(X) = R(X)$. One's first guess might be that the partial sums of the Taylor expansion of $f \in A(X)$ at 0 approximate f; however, it is well-known that these may fail to converge on $\Gamma = \partial\Delta$. On the other hand, the Cesaro means of the Taylor series for f converge uniformly to f , and we may take these as the approximating functions.

Example 9.4. If X is a simply connected set whose boundary is homeomorphic to the unit circle Γ , then $R(X) = A(X)$. This result, the natural generalization of 9.3, was proved by Walsh [96]. The argument, however, is no longer elementary: it depends on theorems from conformal mapping.

Quite clearly, there are two extreme forms X can take: X may be nowhere dense (i.e., $X^o = \emptyset$) or X may be the closure of its interior. Lavrentiev [60] showed that if X is simply connected and nowhere dense then every function in $C(X)$ can be approximated uniformly on X by polynomials. Keldysh [58] proved that if $X = \overline{X^o}$ and X is simply connected then $A(X) = R(X)$, i.e. every function in $A(X)$ is uniformly approximable by polynomials. The case of an arbitrary simply connected compact set remained unsettled until 1951, when Mergelyan [66] proved the beautiful

Theorem 9.5. A necessary and sufficient condition that every function in $A(X)$ be uniformly approximable on X by polynomials is that $S^2 \setminus X$ be connected.

Of course, one half of this theorem is virtually trivial. What concerns us is that if $S^2 \setminus X$ is connected then $A(X) = R(X)$.

If $S^2 \setminus X$ contains more than one component the situation is, quite naturally, more complicated. Mergelyan [67] showed that if $S^2 \setminus X$ has only finitely many components then $A(X) = R(X)$. He also obtained some sufficient conditions for $R(X)$ to coincide with $A(X)$ in the case of infinite connectivity. As evidence of the complexity of the infinitely connected case we submit the following example, due to Mergelyan [67] and known affectionately as Mergelyan's Swiss Cheese.

Example 9.6. Let X be the compact set obtained by removing from $\bar{\Delta}$ a sequence of open discs Δ_n satisfying

(1) $\bar{\Delta}_n \subset \Delta$ for all n ;

(2) $\bar{\Delta}_n \cap \bar{\Delta}_m = \emptyset$, unless $n = m$;

(3) $X = \bar{\Delta} \setminus \bigcup_{n=1}^{\infty} \Delta_n$ has no interior ;

(4) $\sum_{n=1}^{\infty} r_n < \infty$, where r_n is the radius of Δ_n.

One way to construct such a set is to enumerate the points of Δ with rational coordinates and then to proceed in the obvious manner, always choosing $r_n < 2^{-n}$. We claim $C(X) \neq R(X)$. Indeed, let μ be the measure on $\Gamma \cup \bigcup_{n=1}^{\infty} \partial\Delta_n$ which coincides with dz on Γ and with $-dz$ on $\bigcup_{n=1}^{\infty} \partial\Delta_n$.

By (4), μ is a finite measure on X. Cauchy's theorem gives

$$\int_X r(z)\, d\mu(z) = 0$$

for any rational function r with poles off X. Thus

$$\int_X f(z)\, d\mu(z) = 0$$

if $f \in R(X)$. Since μ is clearly a nonzero measure, there exists $g \in C(X)$ such that

$$\int_X g(z)\, d\mu(z) \neq 0 .$$

Hence $R(X) \neq C(X)$.

Thus, there exist compact sets without interior for which $R(X) \neq C(X)$. In the other direction, Hartogs and Rosenthal [44] showed that if X has planar (Lebesgue) measure zero then $R(X) = C(X)$.

The set constructed in 9.6 was nowhere dense. It is natural to ask if $R(X)$ must coincide with $A(X)$ if X is the closure of its interior. This is not the case, even if we make other, quite strong assumptions on X. Before we give an example we need

Lemma 9.7. (Urysohn [85]; Denjoy [22]). Let K be the usual Cantor ternary set on the unit interval. Then there exists a nonzero function g continuous on S^2 and analytic off $K \times K$ such that $g'(\infty) \neq 0$.

Proof. Let $Q = K \times K$. Q is a planar Cantor set and can be constructed (without reference to K) in the following familiar manner. Divide the unit square $0 \leq x,y \leq 1$ into nine equal squares and delete the union of the open squares containing points (x,y) such that $1/3 < x < 2/3$ or $1/3 < y < 2/3$. This process is iterated in the usual way. After the nth step, we are left with 4^n closed squares of total area $(4/9)^n$; call the union of these squares Q_n. Then $Q = \bigcap_{n=1}^{\infty} Q_n$ is a totally disconnected perfect set of planar measure zero. Denote the centers of the components of Q_n by $z_{n,k}$ $(k = 1,2,\ldots,4^n)$, and let

$$g_n(z) = 4^{-n} \sum_{k=1}^{4^n} (z-z_{n,k})^{-1} .$$

It can be shown that $\lim_{n\to\infty} g_n(z) \equiv g(z)$ exists for all z and that g is continuous on S^2. The convergence is uniform off any neighborhood of Q, so that g is analytic off Q and

$$g'(\infty) = \int_{|z|=2} g(z)dz = 2\pi i \neq 0 ,$$

by the residue theorem.

9.7 can also be proved using the more general techniques of Arens [7].

<u>Example</u> 9.8. Let X be the compact set constructed as follows. Delete from the closure of the disc $D = \{|z| < 2\}$ a sequence of open discs Δ_n with radii r_n such that

(1) $\overline{\Delta}_n \subset D \setminus Q$ for all n ;

(2) $\overline{\Delta}_n \cap \overline{\Delta}_m = \emptyset$, unless $m = n$;

(3) $\sum_{n=1}^{\infty} r_n < \infty$;

(4) $\{\Delta_n\}$ accumulates at every point of Q and at no other points.

Then

(a) $X = \overline{X^o}$.

(b) $R(\partial X) = C(\partial X)$.

(c) $\partial X \setminus$ (peak points of $R(X)$) has measure 0 $(dx\, dy)$.

(d) $R(X) \neq A(X)$.

It is clear from the construction that (a) holds. Since

$\partial X = \partial D \cup (\bigcup_{n=1}^{\infty} \partial\Delta_n) \cup Q$ has planar measure 0 , (b) follows from the theorem of Hartogs and Rosenthal quoted above (see 13.6); also, it is clear that every point of $\partial D \cup \bigcup_{n=1}^{\infty} \partial\Delta_n$ is a peak point for $R(X)$ so that ($\partial X \setminus$ peak points of $R(X)) \subset Q$, a set of measure 0 . It remains to show that $R(X) \neq A(X)$. Let μ be the measure on ∂X defined as dz on ∂D , $-dz$ on $\bigcup_{n=1}^{\infty} \partial\Delta_n$, and 0 on Q . By (3), μ is a finite measure on X . As in 9.6 ,

$$\int_X f(z)\, d\mu(z) = 0$$

if $f \in R(X)$, since by Cauchy's theorem μ kills rational functions with poles off X . Let g be the function of 9.7. Then

$$\int_X f(z)\, d\mu(z) = \int_{|z|=2} g(z)dz - \sum_{n=1}^{\infty} \int_{\partial\Delta_n} g(z)dz = 2\pi i ,$$

each term in the infinite sum being equal to 0 by Cauchy's theorem. Thus $g \notin R(X)$. Since $g \in A(X)$, we have $R(X) \neq A(X)$.

Note that 9.8 shows that $R(X)$ and $A(X)$ can have the same peak points except for a set of measure zero and yet fail to coincide. It is easy to modify the construction of 9.8 to insure that $R(X)$ and $A(X)$ fail to have the same peak points. Indeed, choose a point $q \in Q$ that is a peak point for the algebra $A(S^2;Q)$ of all functions continuous on S^2 and analytic off Q. Such a point exists by 2.3 and the maximum modulus principle. (Actually, in order for 2.3 to be applicable it is necessary that $A(S^2; Q)$ separate points. To verify this condition we use an idea of Wermer and choose $z_1, z_2 \in \mathbb{C} \setminus Q$ such that $g(z_1) \neq g(z_2)$. Then the functions $g(z)$, $[g(z)-g(z_1)]/(z-z_1)$, and $[g(z)-g(z_2)]/(z-z_2)$ separate points on S^2). Suppose we remove the Δ_n in such a way that

$$\sum_{n=1}^{\infty} \frac{r_n}{\rho(q,\Delta_n)} < \infty ,$$

where $\rho(q,\Delta_n)$ is the distance from q to Δ_n. Then q cannot be a peak point for $R(X)$ since the measure given by $(2\pi i)^{-1} (z-q)^{-1} dz$ on Γ and $-(2\pi i)^{-1} (z-q)^{-1} dz$ on $\bigcup_{n=1}^{\infty} \Delta_n$ is a finite (complex) measure that represents q on $R(X)$ (cf. the proof of 8.1). But clearly q is a peak point for $A(X)$.

In view of 9.8, it is interesting that the following result is valid.

Theorem 9.9. If $R(X) = A(X)$ then $R(\partial X) = C(\partial X)$.

Proof. If $R(\partial X) \neq C(\partial X)$, then by 2.4 there is a set $E \subset \partial X$

of positive measure such that no point of E is a peak point for $R(\partial X)$; without loss of generality we may assume E is closed. Let

$$g(z) = \int_E \frac{dxdy}{\zeta - z} \qquad \zeta = x + iy .$$

g is quite clearly analytic on $S^2 \setminus E$. Moreover, g is continuous on S^2, since it is the convolution of a locally integrable function and a bounded function of compact support (the characteristic function of E). Finally, $g \not\equiv 0$ because $g'(\infty) = \lim_{z\to\infty} zg(z) = -\int_E dxdy \neq 0$. Consider the algebra $A = A(S^2;E)$ of all functions continuous on S^2 and analytic on $S^2 \setminus E$; since $g \in A$, $A \neq \mathbb{C}$. As above, E contains peak points for A. Since $A \subset A(X)$, E contains peak points for $A(X)$. But since no function in $R(\partial X)$ peaks at any point of E, no function in $R(X)$ can peak on E. Thus $R(X) \neq A(X)$, as required.

Our final (counter-) example is especially curious; it is due to Dolzhenko [28], whose construction was slightly more involved.

Example 9.10. Let J be a Jordan arc of positive planar measure. Let D be an open disc such that $J \subset \overline{D}$ and $J \cap \partial D$ consists of a single point p. Delete from D a sequence of open discs Δ_n with radii r_n such that

(1) $\overline{\Delta}_n \subset D$ for all n ;

(2) $\overline{\Delta}_n \cap \overline{\Delta}_m = \emptyset$ unless $n = m$;

(3) $\overline{\Delta}_n \cap J = \{p_n\}$, a singleton ;

(4) $\sum_{n=1}^{\infty} r_n < \infty$;

(5) $\{\Delta_n\}$ accumulates at each point of J and at no other points.

Clearly, $\overline{X^o} = X$ and, as before, $R(X) \neq A(X)$. (Here we use the fact that the function

$$g(z) = \int_J \frac{dxdy}{\zeta - z} \qquad \zeta = x + iy$$

is in $A(X)$ but not in $R(X)$). What is of special interest in this example is that X^o is simply connected. Thus, the connectivity of the interior of X has little to do with the possibility of approximation.

The examples of this section have indicated the difficulty and complexity of studying rational approximation on arbitrary plane compacta. We now turn our attention to some ideas necessary to Vitushkin's ingenious solution of this problem.

10. AC CAPACITY

In studying the peak points of $R(X)$ we were led to make use of the analytic capacity γ of a set as a measure of its thinness. To study $A(X)$ it is necessary to introduce another such measure, the AC capacity α.

Definition 10.1. Let $S \subset \mathbf{C}$ be arbitrary and $m > 0$. Then $C(S,m)$ is the set of all functions f for which

(1) $f \in C(S^2)$;

(2) $f(\infty) = 0$;

(3) f is analytic off some compact subset of S ;

(4) $\| f \|_\infty \leq m$.

Definition 10.2. $\alpha(S) = \sup_{f \in C(S,1)} |f'(\infty)|$.

This definition differs from that of the analytic capacity γ in that we require not only boundedness of the functions under consideration but also continuity; further, we demand analyticity on all of $S^2 \setminus S$ (not just $\Omega(S)$) .

Proposition 10.3. $\alpha(S) \leq \gamma(S)$.

Proof. This is clear .

Even in the simplest cases the analytic capacity of a set and its AC capacity may fail to coincide. For instance, if $K = [0,1]$ then $\gamma(K) = 1/4$ by 3.9. On the other hand, $\alpha(K) = 0$, since any function analytic on $S^2 \setminus [0,1]$ and continuous on S^2 is analytic on all of S^2 (Morera's theorem) and hence is constant. There is, however, one important case in which the two capacities do give the same result.

Proposition 10.4. Let U be an open set. Then $\alpha(U) = \gamma(U)$.

Proof. Exercise.

AC capacity was introduced by Dolzhenko [28] and is a more natural tool for studying continuous analytic functions than (ordinary) analytic capacity. For instance, the examples of 9.8, 9.9, and 9.10 were based on the fact that the quantities $\alpha(Q)$, $\alpha(E)$, and $\alpha(J)$ did not vanish. Generally speaking, however, AC capacity is considerably less tractable than analytic capacity. Thus, it is not true that $\alpha(K) = \inf_{U \supset K} \alpha(U)$; indeed, this would imply $\alpha(K) = \gamma(K)$ for all compact K via 3.10 and 10.4.

Although we cannot enter into a detailed discussion of the similarities and differences between α and γ - - a careful rereading of section 3 will orient the reader in this direction--we should point out that the following analogues of 3.15 and 5.7 remain valid for functions with continuous boundary behavior.

Proposition 10.5. Let $K \subset \mathbb{C}$ be compact, $f \in C(K,1)$. Then for $z \notin K$

$$|f(z)| \leq \alpha(K)/\rho(z,K) \quad ,$$

where $\rho(z,K)$ is the distance from z to K. If $K \subset \{|z-z_0| \leq R\}$ and

$$f(z) = \frac{a_1}{z-z_0} + \frac{a_2}{(z-z_0)^2} + \dots .$$

then

$$|a_n| \leq e\alpha(K)R^{n-1}\, n \qquad n = 2,3,\dots .$$

Proposition 10.6. Let $K \subset \Delta$ be compact and let $f \in C(\overline{\Delta})$ be analytic on $\Delta \setminus K$. Then

$$\left|\int_\Gamma f(\varsigma)d\varsigma\right| \leq c\alpha(K)\|f\|_\infty ,$$

where c is a universal constant.

Remark. The proof of 10.5 is a simple exercise, while the argument of section 5 (with appropriate modifications) can be used to establish 10.6.

We now turn to a slightly technical result (10.8) which will be useful to us in the sequel. First, we need the following auxiliary

Lemma 10.7. Let p be a positive integer and suppose $0 \leq c_{n,k} \leq 1$ for $k = 1,2,\ldots,np$ $n = 1,2,\ldots$. Then

$$\left(\sum_{n,k} \frac{c_{n,k}}{n}\right)^2 \leq 4p \sum_{n,k} c_{n,k}$$

Proof. Exercise.

Proposition 10.8. Let $S \subset \mathbb{C}$ be given and suppose $\{S_j\}$ is a collection of subsets of S such that each disc of radius $\alpha(S)$ meets at most p of the sets $\{S_j\}$. Then

$$\sum_{j=1}^{\infty} \alpha(S_j) < 400p\, \alpha(S) .$$

Moreover, if $f_j \in C(S_j,1)$ then

$$\max \sum_{j=1}^{\infty} |f_j(z)| < 200p .$$

Proof. Obviously, we may assume $\alpha(S) > 0$. Fix z_0 and renumber the sets S_j using double indices n,k to obtain a double sequence $\{S_{n,k}\}$ such that $n\alpha(S) \leq \rho(z_0,S_{n,k}) < (n+1)\alpha(S)$; here $\rho(z_0,S_{n,k})$ is the distance from z_0 to $S_{n,k}$. By hypothesis, there are at most p sets $S_{o,k}$; moreover, since the annulus $\{n\alpha(S) \leq |z-z_0| \leq (n+1)\alpha(S)\}$ can be covered by 20n discs of radius $\alpha(S)$, there are at most 20np distinct sets $S_{n,k}$ for a fixed value of n . Now let $f_{n,k} \in C(S_{n,k},1)$. Then by 10.5

$$|f_{n,k}(z_0)| \leq \alpha(S_{n,k}) \,/\, \rho(z_0,S_{n,k}) \leq \alpha(S_{n,k}) \,/\, n\alpha(S)$$

for $n \geq 1$. Estimating $|f_{o,k}(z_0)|$ by 1 and applying 10.7 , we have

$$\sum_{j=1}^{\infty} |f_j(z_0)| \leq p + \sum_{n,k} \frac{\alpha(S_{n,k})}{n\alpha(S)} < p + 9\sqrt{pm} ,$$

where $m = \sum_{j=1}^{\infty} \alpha(S_j)/\alpha(S)$. Since z_0 was arbitrary,

$$\max \sum_{j=1}^{\infty} |f_j(z)| \leq p + 9\sqrt{pm} .$$

Now take $\varphi_j \in C(S_j,1)$ such that $\varphi_j'(\infty) = (1/2)\alpha(S_j)$ and let $\varphi = \sum \varphi_j$. Then $\|\varphi\|_\infty \le p + 9\sqrt{pm}$ so that

$$\varphi'(\infty) = \frac{1}{2}\sum_{j=1}^{\infty} \alpha(S_j) \le \alpha(S)\ [p + 9\sqrt{pm}] ,$$

whence $m \le 2p + 18\sqrt{pm}$. Then $m < 400p$, which is the first inequality. It follows that

$$\max \sum_{j=1}^{\infty} |f_j(z)| \le p + 9\sqrt{400p^2} < 200p ,$$

as required. We are done.

Now suppose that $S \subset \mathbb{C}$ is a bounded set for which $\alpha(S) > 0$. Then if $f \in C(S,m)$ and $z_0 \in \mathbb{C}$ we have

$$f(z) = \frac{a_1}{z-z_0} + \frac{a_2}{(z-z_0)^2} + \ldots .$$

Set

$$\beta(S,z_0,f) = \frac{a_2}{\alpha(S)} .$$

Clearly,

$$\beta(S,z_0,f) = \frac{1}{\alpha(S)} \frac{1}{2\pi i} \int_\sigma f(\zeta)(\zeta-z_0)d\zeta ,$$

where σ is a rectifiable contour having winding number 1 about each point of S (for instance, a large circle). Set

$$\beta(S,z) = \sup |\beta(S,z,f)| ,$$

where the sup is taken over all $f \in C(S,1)$. Finally, let

$$\beta(S) = \inf_z \beta(S,z) .$$

It is easy to see that $\beta(S,z)$ is a continuous function of z on $\mathbb{C}$; moreover, a simple computation shows that $\beta(S,z) \to \infty$ as $z \to \infty$. Hence, there exists some point $O(S)$ such that $\beta(S,O(S)) = \beta(S)$.

We shall not develop the properties of β in a systematic fashion; the interested reader is invited to pursue this line of thought on his own. However, there are two basic observations that should be made.

<u>Proposition</u> 10.9. $\alpha(S) \leq \beta(S)$.

<u>Proof</u>. Let $\varphi \in C(S,1)$ with

$$\varphi(z) = \frac{a_1}{z-O(S)} + \frac{a_2}{(z-O(S))^2} + \dots ,$$

where $|a_1| \geq \alpha(S) - \varepsilon$. Then

$$\beta(S) = \beta(S,O(S)) \geq |\beta(S,O(S),\varphi^2)| \geq \frac{(\alpha(S)-\varepsilon)^2}{\alpha(S)} .$$

Let $\varepsilon \to 0$ to obtain the desired result.

Proposition 10.10. Let α,β be complex numbers such that $|\alpha| \leq \alpha(S)$ and $|\beta| \leq \beta(S)$. Then there exists $g \in C(S,20)$ such that $g'(\infty) = \alpha$ and $\beta(S,O(S),g) = \beta$.

Proof. It follows from the integral formula for $\beta(S,z,f)$ that

$$\beta(S,z,f) = \beta(S,t,f) + \frac{f'(\infty)}{\alpha(S)} [t-z] .$$

Take $\varphi \in C(S,2)$ such that $\varphi'(\infty) = \alpha(S)$ and let $z_0 = O(S) + \beta(S,O(S),\varphi)$. Then

$$\beta(S,z_0,\varphi) = \beta(S,O(S),\varphi) + \frac{\varphi'(\infty)}{\alpha(S)} [O(S)-z_0] = 0 .$$

Next choose $f_1 \in C(S,2)$ such that $\beta(S,z_0,f_1) = \beta(S)$. Finally, take ε so that $f_1'(\infty) + \varepsilon\,\varphi'(\infty) = 0$ and set $f_2 = f_1 + \varepsilon\varphi$. Then since $f_2'(\infty) = 0$ we have

$$\beta(S,0(S),f_2) = \beta(S,z_0,f_2) = \beta(S,z_0,f_1) + \varepsilon\,\beta(S,z_0,\varphi)$$

$$= \beta(S,z_0,f_1) = \beta(S) .$$

Also, since $f_1 \in C(S,2)$, $|\varepsilon| \leq 2$, so that $f_2 \in C(S,6)$. Recalling that $\beta(S,0(S),\varphi) = z_0 - 0(S)$, we have $|0(S)-z_0| \leq 2\beta(S)$. Then

$$f = \frac{\alpha}{\alpha(S)}\,\varphi + \frac{\alpha[0(S)-z_0]}{\alpha(S)\beta(S)}\,f_2 + \frac{\beta}{\beta(S)}\,f_2 \in C(S,20)$$

is the function we seek .

It is easy to see that the constant 20 of 10.10 can be improved considerably; indeed, by a more careful choice of the norms of φ and f_1 we can insure that $f \in C(S,5+\delta)$, where $\delta > 0$ is arbitrary. It would be interesting to know if and by how much the constant 5 can be improved.

11. A SCHEME FOR APPROXIMATION

This section deals with constructions and estimates which will enable us to prove Vitushkin's theorem. It also contains (11.8) a constructive proof, due to Garnett, of Bishop's celebrated localization theorem.

Let $K(z_o,\delta) = \{|z-z_o| \leq \delta\}$, $K^o(z,\delta) = \{|z-z_o| < \delta\}$. If $f \in C(S^2)$, its modulus of continuity $\omega_f(\delta)$ is given by $\omega_f(\delta) = \sup_{|x-y|<\delta} |f(x)-f(y)|$. If $V \subset \mathbf{C}$, we shall write CV for $S^2 \setminus V$. Finally, recall that the operator $\frac{\partial}{\partial\bar{z}}$ is given by $\frac{\partial}{\partial\bar{z}} = \frac{1}{2}\left(\frac{\partial}{\partial x} + i\,\frac{\partial}{\partial y}\right)$, where $z = x + iy$.

<u>Proposition</u> 11.1. Let $\delta_n \downarrow 0$. For each positive integer n there exists a C^∞ partition of unity $\{\varphi_{k,n}\}$ which satisfies

(1) $0 \leq \varphi_{k,n}(z) \leq 1$, $\varphi_{k,n} \in C_R^\infty(S^2)$.

(2) $\sum\limits_{k=1}^{\infty} \varphi_{k,n} \equiv 1$.

(3) $\varphi_{k,n}(z) = 0$ off $K(z_{k,n},\delta_n)$.

(4) No point z is contained in more than 25 of the discs $K(z_{k,n},\delta_n)$.

(5) $\| \frac{\partial\varphi_{k,n}}{\partial\bar{z}}\|_\infty \leq 20/\delta_n$.

Proof. Take $\varphi \in C_R^\infty(S^2)$ such that $0 \leq \varphi(z) \leq 1$, $\varphi(z) = 0$ for $|z| \geq 1$, $\iint \varphi(x+iy)dxdy = 1$, and $\| \frac{\partial\varphi}{\partial\bar{z}} \|_\infty \leq 10$. Divide the plane into unit squares Q_k , $k = 1,2,3,\ldots,$ having mutually disjoint interiors and set

$$\varphi_k(z) = \iint_{Q_k} \varphi(\zeta - z)\, dxdy \qquad (\zeta = x+iy) .$$

The functions φ_k clearly satisfy (1) , and

$$\sum_{k=1}^{\infty} \varphi_k(z) = \sum_{k=1}^{\infty} \iint_{Q_k} \varphi(\zeta-z)dxdy = \iint \varphi(\zeta-z)dxdy = 1 ,$$

so that (2) holds. Moreover, if z lies at a distance of at least 1 from Q_k then $\varphi_k(z) = 0$; thus each φ_k is supported on a disc $K(z_k,2)$, where z_k is the centroid of Q_k . It follows that no point z belongs to more than 25 of the $K(z_k,2)$.

Now fix n and set $\varphi_{k,n}(z) = \varphi_k(2z/\delta_n)$. Then $\varphi_{k,n}$ is supported on a disc $K(z_{k,n}, \delta_n)$ and no point z belongs to more than 25 of the $K(z_{k,n}, \delta_n)$. Clearly, $\| \frac{\partial \varphi_{k,n}}{\partial \bar{z}} \|_\infty \leq 10 \cdot \frac{2}{\delta_n}$. **Finally,** the $\varphi_{k,n}$ satisfy (1) and (2) because the φ_k do. This completes the proof.

If f is a continuous function on S^2 and g is a continuously differentiable function of compact support we shall write

$$f_g(z) = \frac{1}{\pi} \iint \frac{f(\zeta)-f(z)}{\zeta - z} \frac{\partial g}{\partial \bar{\zeta}} \, dxdy$$

$$= f(z)g(z) + \frac{1}{\pi} \iint f(\zeta) \frac{\partial g}{\partial \bar{\zeta}} \frac{1}{\zeta - z} \, dxdy \qquad (\zeta = x+iy) .$$

(We have used Green's formula for g in passing to the second equality). Let a_f be the set of points at which f is analytic and s_g the closed support of g. Let $n_g = S^2 \setminus s_g$. Then we have

Proposition 11.2. Let f and g be as above and suppose that the diameter of s_g does not exceed δ. Then

(1) $f_g \in C(S^2)$, and $f_g(\infty) = 0$.

(2) f_g is analytic on $a_f \cup n_g$.

(3) $\| f_g \|_\infty \leq 2\delta\omega_f(\delta) \| \frac{\partial g}{\partial \bar{z}} \|_\infty$

<u>Proof</u>. (1) $f_g \in C(S^2)$ since $f_g - gf$ is the convolution of a bounded function of compact support and a locally integrable function. Moreover, it is clear that $f_g(\infty) = 0$. (2) can be established by a straightforward calculation, which we leave to the reader. For (3), take z_0 such that $|f_g(z_0)| = \| f_g \|_\infty$. Then

$$\| f_g \|_\infty = \left| \frac{1}{\pi} \iint \frac{f(\zeta)-f(z_0)}{\zeta - z_0} \frac{\partial g}{\partial \bar{\zeta}} \, dxdy \right|$$

$$\leq \frac{1}{\pi} \omega_f(\delta) \| \frac{\partial g}{\partial \bar{\zeta}} \|_\infty \iint_{S_g} \frac{dxdy}{|\zeta - z_0|} .$$

Since

$$\iint_{S_g} \frac{dxdy}{|\zeta - z_0|} \leq \iint_{K(0,\delta)} \frac{dxdy}{|\zeta|} = 2\pi\delta ,$$

we are done.

Our next order of business is a coefficient estimate (11.6). The proof, which is quite involved, will be presented in a series of lemmas.

<u>Lemma</u> 11.3. Let $f \in A(X)$ and let φ be a continuously differentiable function supported on $K(z_0, \delta)$. Then

$$\left|\frac{1}{\pi}\iint f(\zeta)\frac{\partial\varphi}{\partial\bar\zeta}\,dxdy\right| \leq 4\delta\cdot\omega_f(2\delta)\cdot\left\|\frac{\partial\varphi}{\partial\bar z}\right\|_\infty\cdot\alpha(CX^o\cap K(z_o,\delta)).$$

Proof. We may assume that $f\in C(S^2)$. By (3) of 11.2, $\|f_\varphi\|_\infty \leq 4\delta\cdot\omega_f(2\delta)\cdot\left\|\frac{\partial\varphi}{\partial\bar z}\right\|_\infty$. Moreover, f_φ is analytic on $X^o\cup CK(z_o,\delta)$ so that, by the definition of α, we have

$$|f_\varphi{}'(\infty)| \leq 4\delta\cdot\omega_f(2\delta)\cdot\left\|\frac{\partial\varphi}{\partial\bar z}\right\|_\infty\cdot\alpha(CX^o\cap K(z_o,\delta)).$$

But clearly

$$f_\varphi{}'(\infty) = \lim_{z\to\infty} zf_\varphi(z) = -\frac{1}{\pi}\iint f(\zeta)\frac{\partial\varphi}{\partial\bar\zeta}\,dxdy .$$

We are done .

Lemma 11.4. Let f and φ be as in 11.3. Then

$$\left|\frac{1}{\pi}\iint f(\zeta)\cdot\frac{\partial\varphi}{\partial\bar\zeta}\cdot(\zeta-z_o)dxdy\right| \leq 4\delta^2\omega_f(2\delta)\left\|\frac{\partial\varphi}{\partial\bar z}\right\|_\infty\alpha(CX^o\cap K(z_o,\delta)).$$

Proof. Let $g(\zeta)=(\zeta-z_o)\varphi(\zeta)$. Then $\frac{\partial g}{\partial\bar\zeta}=(\zeta-z_o)\frac{\partial\varphi}{\partial\bar\zeta}$, so that $\left\|\frac{\partial g}{\partial\bar\zeta}\right\|_\infty\leq\delta\left\|\frac{\partial\varphi}{\partial\bar\zeta}\right\|_\infty$. Applying 11.3 with φ replaced by g, we obtain the desired result.

Now let the compact set X be fixed. Suppose that there

exist constants $m \geq 1$, $r \geq 1$ such that for all z and all $\delta > 0$

$$(*) \qquad \alpha(CX^{o} \cap K(z,\delta)) \leq m\alpha(CX \cap K(z,r\delta)) .$$

Let $\{\varphi_{k,n}\}$ be the sequence of partitions of unity constructed in 11.1 and set $X_{k,n} = CX \cap K(z_{k,n}, r\delta_n)$. We have

<u>Lemma</u> 11.5. Suppose X satisfies (*) for all z and all $\delta > 0$, and let $f \in A(X)$. Then

$$\left|\frac{1}{\pi} \iint f(\zeta) \frac{\partial \varphi_{k,n}}{\partial \bar{\zeta}} (\zeta - \mathcal{O}(X_{k,n}))dxdy\right| \leq m_1 \cdot \omega_f(2\delta_n) \cdot \alpha(X_{k,n}) \cdot \beta(X_{k,n}) ,$$

where m_1 is a constant depending only on m and r .

<u>Proof.</u> For convenience of notation we shall write $\varphi = \varphi_{k,n}$, $\delta = \delta_n$, $z_o = z_{k,n}$, $\Theta = \mathcal{O}(X_{k,n})$. Choose β so that

$$(1) \qquad \begin{cases} \beta \leq \beta(X_{k,n}) \leq 2\beta & \text{if} \quad \beta(X_{k,n}) < \delta \\ \beta = \delta & \text{if} \quad \beta(X_{k,n}) \geq \delta \end{cases}$$

and cover $K(z_o, \delta)$ by (a finite number of) discs $K(t_i, \beta)$ in such a way that each disc $K(z,\beta)$ of radius β intersects at most $25r^2$ of the discs $K(t_i, r\beta)$ and $K(t_i, r\beta) \subset K(z_0, r\delta)$ for all i. By (1) and 10.8 we have $\alpha(X_{k,n}) \leq (2+r)\beta$. It follows easily (cf. the proof of 10.8) that each disc of radius $\alpha(X_{k,n})$ meets at most p of the discs $K(t_i, r\beta)$, where p is a constant depending only on r. Choose C^∞ functions g_i supported on $K(t_i, \beta)$ such that $0 \leq g_i \leq 1$, $\sum_i g_i(z) = 1$ on a neighborhood of $K(z_0, \delta)$, and $\|\frac{\partial g_i}{\partial \bar{z}}\|_\infty \leq 20/\beta$. Then

$$(2) \quad \frac{1}{\pi} \iint f(\zeta) \frac{\partial \varphi}{\partial \bar{\zeta}} (\zeta - \sigma)\, dxdy =$$

$$\sum_i \frac{1}{\pi} \iint f(\zeta) \frac{\partial}{\partial \bar{\zeta}} (\varphi g_i)(\zeta - t_i)\, dxdy + \sum_i \frac{1}{\pi} \iint f(\zeta) \frac{\partial}{\partial \bar{\zeta}} (\varphi g_i)(t_i - \sigma)\, dxdy$$

Since $\beta \leq \delta$, we have by 11.3

$$(3) \quad \left|\frac{1}{\pi}\iint f(\zeta) \frac{\partial}{\partial \bar{\zeta}}(\varphi g_i)\, dxdy\right| \leq 4\beta \cdot \omega_f(2\beta) \cdot \left(\frac{20}{\delta} + \frac{20}{\beta}\right) \cdot \alpha(CX^o \cap K(t_i, \beta))$$

$$\leq 160 \cdot \omega_f(2\beta) \cdot \alpha(CX^o \cap K(t_i, \beta)) .$$

Similarly, by 11.4 we have

(4) $\left|\frac{1}{\pi}\iint f(\zeta)\frac{\partial}{\partial\bar\zeta}(\omega g_i)(\zeta-t_i)dxdy\right| \leq 160\beta \cdot \omega_f(2\beta) \cdot \alpha(CX^o \cap K(t_i,\beta)).$

It follows from (2), (3), and (4) that

(5) $\left|\frac{1}{\pi}\iint f(\zeta)\frac{\partial\varphi}{\partial\bar\zeta}(\zeta-\theta)dxdy\right| \leq$

$$160\beta \cdot \omega_f(2\beta) \cdot \sum_i \alpha(CX^o \cap K(t_i,\beta)) \quad +$$

$$160\omega_f(2\beta) \sum_i |t_i-\theta| \ \alpha(CX^o \cap K(t_i,\beta)) \ .$$

Since each disc of radius $\alpha(X_{k,n})$ meets at most p of the discs $K(t_i,r\beta) \subset K(z_o,r\delta)$, it follows from (*) and 10.8 that

$$\sum_i \alpha(CX^o \cap K(t_i,\beta)) \leq m \sum_i \alpha(CX \cap K(t_i,r\beta))$$

$$\leq m \cdot 400p \cdot \alpha(X_{k,n}) \ .$$

Accordingly, by (1) ,

(6) $160\beta \cdot \omega_f(2\beta) \sum_i \alpha(CX^o \cap K(t_i,\beta)) \leq C_1 \cdot mp \cdot \beta \cdot \omega_f(2\beta)\alpha(X_{k,n})$

$$\leq C_1 \cdot mp \cdot \omega_f(2\beta)\alpha(X_{k,n})\beta(X_{k,n}),$$

which is the bound required for the first term on the r.h.s. of (5). (Here C_1 is an absolute constant) .

Estimating the second sum in (5) is more complicated. For $t_i \neq \theta$, choose $\Psi_i \in C(CX \cap K(t_i,r\beta),1)$ such that

$$(7)\quad \Psi'_i(\infty) = \frac{1}{2}\,\frac{|t_i-\theta|}{t_i-\theta}\cdot \alpha(CX \cap K(t_i,r\beta))$$

and let $\Psi = \sum_i \Psi_i$. By 10.8 $\|\Psi\|_\infty \leq 200p$. Let $\sigma = \partial K(z_0,r\delta)$, $\sigma_i = \partial K(t_i,r\beta)$. Then since $\beta(X_{k,n}) = \beta(X_{k,n},\theta) \geq |\beta(X_{k,n},\theta,\Psi/200p)|$, we have

$$(8)\quad \beta(X_{k,n}) \geq [2\pi\cdot 200p\cdot \alpha(X_{k,n})]^{-1}\left|\int_\sigma \Psi(z)(z-\theta)dz\right|$$

$$\geq [400\pi\cdot p\cdot \alpha(X_{k,n})]^{-1}\left|\sum_i (t_i-\theta)\int_{\sigma_i}\Psi_i(z)dz + \sum_i\int_{\sigma_i}\Psi_i(z)(z-t_i)dz\right| .$$

By (7) ,

$$(9)\quad \sum_i (t_i-\theta)\int_{\sigma_i}\Psi_i(z)dz = \sum_i \frac{1}{2}|t_i-\theta|\alpha(CX \cap K(t_i,r\beta)) ,$$

while 10.5 gives

$$(10)\quad \left|\int_{\sigma_i} \Psi_i(z)(z-t_i)dz\right| \leq 2\pi\cdot 2e\cdot r\beta\cdot \alpha(CX \cap K(t_i,r\beta)) .$$

It follows from (8), (9), and (10) that

$$(11)\quad \sum_i |t_i-\sigma|\ \alpha(CX \cap K(t_i,r\beta)) \leq 800\pi\cdot p\cdot \alpha(X_{k,n})\beta(X_{k,n})$$

$$+\ 8\pi e r\beta \sum_i \alpha(CX \cap K(t_i,r\beta)).$$

Applying 10.8 once again we obtain

$$(12)\quad \sum_i |t_i-\sigma|\ \alpha(CX \cap K(t_i,r\beta)) \leq C_2\cdot rp\cdot \alpha(X_{k,n})\beta(X_{k,n}) ,$$

so that by (*) and (12)

$$(13)\quad 160\omega_f(2\beta) \sum |t_i-\sigma|\ \alpha(CX^\circ \cap K(t_i,\beta))$$

$$\leq 160\ \omega_f(2\beta)\cdot m\cdot \sum |t_i-\sigma|\ \alpha(CX \cap K(t_i,r\beta))$$

$$\leq C_3\cdot rpm\cdot \omega_f(2\beta)\cdot \alpha(X_{k,n})\beta(X_{k,n}) ,$$

where C_3 is an absolute constant. Substituting (6) and (13) into (5) gives

$$(14)\quad \left|\frac{1}{\pi}\iint f(\zeta)\,\frac{\partial\varphi}{\partial\bar\zeta}\,(\zeta-\sigma)\ dxdy\right| \leq$$

$$\leq C_1 \cdot mp \cdot \omega_f(2\beta)\alpha(X_{k,n})\beta(X_{k,n}) + C_3 \cdot mpr \cdot \omega_f(2\beta)\alpha(X_{k,n})\beta(X_{k,n}) .$$

Taking $m_1 = mp(C_1 + rC_3)$ and noting that (1) implies $\omega_f(2\beta) \leq \omega_f(2\delta)$, we obtain the desired result.

Proposition 11.6. Let f , X , and $X_{k,n}$ be as in 11.5 and let

$$g(z) = f(z)\,\varphi_{k,n}(z) + \frac{1}{\pi}\iint f(\zeta)\,\frac{\partial\varphi_{k,n}}{\partial\bar{\zeta}}\,\frac{1}{\zeta - z}\,dxdy$$

$$= \sum_{s=1}^{\infty} \frac{a_s}{(z - O(X_{k,n}))^s} .$$

Then $|a_2| \leq m_1\,\omega_f(2\delta_n)\alpha(X_{k,n})\beta(X_{k,n})$

Proof. As before, we shall write $\varphi = \varphi_{k,n}$, $z_0 = z_{k,n}$, etc. Then for $\sigma = \partial K(z_0, 2\delta)$

$$a_2 = \frac{1}{2\pi i}\int_\sigma g(z)(z-\theta)dz$$

$$= \frac{1}{2\pi i}\int_\sigma \left(\frac{1}{\pi}\iint f(\zeta)\,\frac{\partial\varphi}{\partial\bar{\zeta}}\,\frac{1}{\zeta - z}\,dxdy\right)(z-\theta)dz$$

$$= \frac{1}{\pi}\iint f(\zeta)\,\frac{\partial\varphi}{\partial\bar{\zeta}}\left(\frac{1}{2\pi i}\int_\sigma \frac{z-\theta}{\zeta - z}\,dz\right) dxdy$$

$$= \frac{1}{\pi}\iint f(\zeta)\,\frac{\partial\varphi}{\partial\bar{\zeta}}\,(\theta - \zeta)\,dxdy .$$

By 11.5, we are done.

Finally, we need the following result.

Lemma 11.7. Let X be a compact set and U a bounded open set. Suppose that $f \in R(X)$ and f is analytic on (a neighborhood of) CU. Then $f \in R(X \cup CU)$.

We shall obtain 11.7 as a corollary to a significantly more general result that is of considerable interest in itself, namely, the fact that a function "locally in" $R(X)$ actually belongs to $R(X)$. More precisely, we have

Theorem 11.8. (Bishop). Let $X \subset \mathbb{C}$ be compact, $f \in C(S^2)$. Suppose that for each $z \in X$ there exists a closed neighborhood $K_z = K(z, \delta_z)$ such that $f|_{X \cap K_z} \in R(X \cap K_z)$. Then $f \in R(X)$.

Proof (Garnett). By compactness, we can take $K_{z_1}, \ldots, K_{z_n}$ such that $\bigcup_{j=1}^{n} K_{z_j} \supset X$. Let $K_j = K_{z_j}$. Choose $\varphi_j \in C^\infty(S^2)$ $(j = 1,2,\ldots,n)$ such that $0 \leq \varphi_j(z) \leq 1$, $\varphi_j = 0$ off K_j, and $\varphi(z) = \sum \varphi_j(z) = 1$ for $z \in V$, where V is a closed neighborhood of X contained in $\bigcup_{j=1}^{n} K_j$. Then if

$$f_j(z) = f(z)\varphi_j(z) + \frac{1}{\pi} \iint f(\zeta) \frac{\partial \varphi_j}{\partial \bar{\zeta}} \frac{1}{\zeta - z} \, dxdy$$

we have $\sum_{j=1}^{n} f_j = f + \Phi$, where $\Phi(z) = \frac{1}{\pi} \iint f(\zeta) \frac{\partial \varphi}{\partial \bar{\zeta}} \frac{1}{\zeta - z} \, dxdy$.

Now let $\varepsilon > 0$ be given, and set

$$C = \max_{1 \le j \le n} \sup_{z} \frac{1}{\pi} \iint \left| \frac{\partial \varphi_j}{\partial \bar{\zeta}} \right| \frac{1}{|\zeta - z|} \, dxdy .$$

Choose rational functions h_j , $1 \le j \le n$, with poles off X such that $\| f - h_j \|_{X \cap K_j} < \varepsilon/8nC$. Then on a suitable closed neighborhood U_j of $X \cap K_j$ we have $\| f - h_j \|_{U_j} < \varepsilon/4nC$ (and h_j is analytic on U_j) . Modify h_j off U_j in such a way that $\| f - h_j \|_{S^2} < \varepsilon/4nC$ and set

$$g_j(z) = h_j(z)\varphi_j(z) + \frac{1}{\pi} \iint h_j(\zeta) \frac{\partial \varphi_j}{\partial \bar{\zeta}} \frac{1}{\zeta - z} \, dxdy$$

By 11.2 , g_j is analytic on $U_j \cup CK_j$, which is a neighborhood of X . Moreover ,

$$\| g_j - f_j \|_{S^2} \le \| h_j - f \|_{S^2} + \left\| \frac{1}{\pi} \iint_{K_j} [h_j(\zeta) - f(\zeta)] \frac{\partial \varphi_j}{\partial \bar{\zeta}} \frac{1}{\zeta - z} \, dxdy \right\|_{S^2}$$

$$\le \| h_j - f \|_{S^2} + \| h_j - f \|_{K_j} \cdot C < \varepsilon/2n .$$

Setting $g = \sum g_j$, we have

$$\|g-\sum f_j\|_V \le \sum_{j=1}^{n} \|g_j-f_j\|_V < \varepsilon/2 .$$

Now since $\frac{\partial\varphi}{\partial\bar{\zeta}} = 0$ on V, $\Phi(z)$ is analytic on V. Therefore, by 9.1, we can find a rational function $r \in R(X)$ such that $\|\Phi-r\|_X < \varepsilon/4$. Moreover, since each g_j is analytic in a neighborhood of X, g is also, so we can find another rational function $q \in R(X)$ satisfying $\|g-q\|_X < \varepsilon/4$. Then $\|f-(r+q)\| < \varepsilon$, as required.

The function Φ can be eliminated in the argument above by modifying f (off X) to have compact support. For then, if V contains the support of f it is easy to see that $\Phi = 0$. For another proof of 11.8 and a comment on its origin see 15.9.

12. VITUSHKIN'S THEOREM

We now have at our disposal the tools necessary to prove Vitushkin's theorem.

Theorem 12.1. (Vitushkin [94]). Let $X \subset \mathbb{C}$ be compact. Then the following are equivalent

(1) $R(X) = A(X)$.

(2) $\alpha(CX \cap G) = \alpha(CX^{o} \cap G)$ for every bounded open set G .

(3) $\alpha(CX \cap D) = \alpha(CX^{o} \cap D)$ for every open disc D .

(4) For each $z \in CX^{o}$ there exists $r \geq 1$ such that

$$\overline{\lim_{\delta \to 0}} \frac{\alpha(CX^{o} \cap K(z,\delta))}{\alpha(CX \cap K(z,r\delta))} < \infty .$$

(5) There exists $r \geq 1$, $m \geq 1$ such that for every z and all δ

$$\alpha(CX^{o} \cap K(z,\delta)) \leq m\alpha(CX \cap K(z,r\delta)).$$

Proof. We shall prove the following chain of implications: (5) => (1) => (2) => (3) => (4) => (5) .

(5) => (1) . Let $f \in A(X)$ and extend f to S^2 as a function of

compact support. Fix n and let $\{\varphi_{k,n}\}$ be the partition of unity constructed in 11.1. For convenience of notation we shall suppress the subscript n in the argument that follows, so that $\delta = \delta_n$, $\varphi_k = \varphi_{k,n}$, $z_k = z_{k,n}$, etc. Then

$$f_k(z) = f(z)\varphi_k(z) + \frac{1}{\pi}\iint f(\zeta)\,\frac{\partial\varphi_k}{\partial\bar{\zeta}}\,\frac{1}{\zeta - z}\,dxdy$$

is analytic on $X^o \cup CK(z_k,\delta)$ and $\sum f_k \equiv f$. Let $X_k = CX \cap K(z_k, r\delta)$. Put $O_k = O(X_k)$ and write

$$f_k(z) = \sum_{s=1}^{\infty} \frac{a_{sk}}{(z-O_k)^s} \quad .$$

By (3) of 11.2 and (5) of 11.1 we have $\|f_k\|_\infty \le 2\cdot 2\delta \cdot \omega_f(2\delta)\cdot 20/\delta = 80\,\omega_f(2\delta)$ so that

$$|a_{1k}| = |f_k'(\infty)| \le 80\omega_f(2\delta)\alpha(CX^o \cap K(z_k,\delta)) \le 80m\omega_f(2\delta)\alpha(X_k),$$

by (5). Also, by 11.6,

$$|a_{2k}| \le m_1\cdot\omega_f(2\delta)\alpha(X_k)\beta(X_k) \ .$$

Let $m_2 = 20\max(80m, m_1)$. According to 10.10 there exists $g_k \in C(X_k, m_2\omega_f(2\delta))$ such that $g_k'(\infty) = f_k'(\infty)$ and

$\beta(X_k, O_k, g_k) = \beta(X_k, O_k, f_k)$. Of course, $\|f_k\|_\infty \leq m_2\omega_f(2\delta)$ so that

$$\|f_k - g_k\|_\infty \leq 2m_2\omega_f(2\delta) \;;$$

hence

$$\frac{|z-z_k|^3}{(r\delta)^3}\,|f_k(z)-g_k(z)| \leq 2m_2\omega_f(2\delta)\,, \qquad |z-z_k| = r\delta\,.$$

Therefore, by the maximum modulus principle,

$$|f_k(z)-g_k(z)| \leq 2m_2 r^3\delta^3\omega_f(2\delta)/|z-z_k|^3 \qquad |z-z_k| \geq r\delta\,.$$

It follows that

$$|f_k(z)-g_k(z)| \leq m_3\omega_f(2\delta)\,\min\left\{1, \frac{\delta^3}{|z-z_k|^3}\right\}\,,$$

where $m_3 = 2m_2r^3$.

Now fix z^*. Each disc $K(z_k,\delta)$ intersects at least one and at most two of the circles $C_s = \{|z-z^*| = s\delta\}$ $(s = 1,2,\ldots.)$. Let N_s be the number of discs which touch C_s. Then by (4) of 11.1, $N_s \cdot \pi\delta^2 \leq 25 \cdot$ area of $\{(s-1)\delta \leq |z-z^*| \leq (s+1)\delta\}$, so that $N_s \leq 100s$. Also, if $s \geq 2$ and $K(z_k,\delta) \cap C_s \neq \emptyset$, we have $|z^*-z_k| \geq (s-1)\delta$, whence

$$|f_k(z^*)-g_k(z^*)| \leq m_3\omega_f(2\delta)\,\frac{1}{(s-1)^3} \ .$$

Therefore,

$$\sum_{k=1}^{\infty}|f_k(z^*)-g_k(z^*)| \leq m_3\omega_f(2\delta)N_1 + m_3\omega_f(2\delta)\sum_{s=2}^{\infty}\frac{N_s}{(s-1)^3}$$

$$\leq m_4\omega_f(2\delta) \ ,$$

where $m_4 = 100m_3\ (1+\sum_{s=2}^{\infty}\frac{s}{(s-1)^3})$. Thus

$$\|f - \sum_{k=1}^{\infty} g_k\|_{S^2} = \|\sum_{k=1}^{\infty} f_k - \sum_{k=1}^{\infty} g_k\|_{S^2} = \|\sum_{k=1}^{\infty}(f_k-g_k)\|_{S^2}$$

$$\leq \sup_z \sum_{k=1}^{\infty}|f_k(z)-g_k(z)| \leq m_4\omega_f(2\delta) \ .$$

Let $\sum'$ denote the (finite) sum over those indices k for which f_k is nonzero. Then $\sum' f_k = f$ and we have $\|\sum'(f_k-g_k)\|_X \leq m_4\omega_f(2\delta)$ Also, each g_k is analytic on a neighborhood of $CX_k = X \cup CK(z_k, r\delta)$; thus $g = \sum' g_k$ is analytic in a neighborhood of X . By 9.1 , $g \in R(X)$.

Since m_4 depends only on m_1, which is fixed, and $\delta = \delta_n \to 0$ as $n \to \infty$, we are done.

(1) => (2). Suppose $R(X) = A(X)$, and let $\varepsilon > 0$ be given. Choose a closed set $K \subset CX^o \cap G$ and a function $f \in C(K,1)$ such that $f'(\infty) > \alpha(CX^o \cap G) - \varepsilon$. Clearly, $f \in A(X)$, so $f \in R(X)$ by (1). Also, f is analytic on a neighborhood of CG; hence $f \in R(X \cup CG)$ by 11.7. Choose rational functions f_n such that (after suitable modifications on $CX \cap G$) $f_n \in C(CX \cap G, 1)$ and $f_n \to f$ uniformly on $X \cup CG$; then $f_n'(\infty) \to f'(\infty)$. Since $|f_n'(\infty)| \leq \alpha(CX \cap G)$, we have $\alpha(CX^o \cap G) - \varepsilon \leq f'(\infty) \leq \alpha(CX \cap G)$. Letting $\varepsilon \to 0$ we get $\alpha(CX^o \cap G) \leq \alpha(CX \cap G)$. The opposite inequality is obvious.

(2) => (3) is trivial.

(3) => (4). Fix $r > 1$. Then

$$\frac{\alpha(CX^o \cap K(z,\delta))}{\alpha(CX \cap K(z,r\delta))} \leq \frac{\alpha(CX^o \cap K^o(z,(\frac{1+r}{2})\delta))}{\alpha(CX \cap K^o(z,(\frac{1+r}{2})\delta))} = 1 .$$

Letting $\delta \to 0$ we obtain

$$\overline{\lim_{\delta\to 0}} \frac{\alpha(CX^o \cap K(z,\delta))}{\alpha(CX \cap K(z,r\delta))} \leq 1$$

for <u>every</u> $r > 1$.

(4) => (5) . Suppose (5) does not hold. Choose z_1 and δ_1 in such a way that $\alpha(CX^o \cap K(z_1,\delta_1)) > \alpha(CX \cap K(z_1,3\delta_1))$. Then $\alpha(CX^o \cap K^o(z_1,2\delta_1)) \neq \alpha(CX \cap K^o(z_1,2\delta_1))$, so that $A(X \cap K(z_1,2\delta_1)) \neq R(X \cap K(z_1,2\delta_1))$ by the fact that (1) => (2). Set $X_1 = X \cap K(z_1,2\delta_1)$. The implication (5) => (1) guarantees the existence of z_2,δ_2 such that $\alpha(CX_1{}^o \cap K(z_2,\delta_2)) > 2\alpha(CX_1 \cap K(z_2,5\delta_2))$. Now $K(z_2,5\delta_2) \subset K(z_1,2\delta_1 + \delta_2)$, for otherwise $CX_1 \cap K(z_2,5\delta_2)$ contains a disc of diameter greater than δ_2 so that $2\alpha(CX_1 \cap K(z_2,5\delta_2)) > \delta_2 \geq \alpha(CX_1^o \cap K(z_2,\delta_2))$, a contradiction. Also, $K(z_2,2\delta_2) \subset K^o(z_1,2\delta_1)$. For otherwise, there exists $x \in K(z_2,2\delta_2) \setminus K^o(z_1,2\delta_1)$ and $x + 2e^{i\theta}\delta_2 \in K(z_2,5\delta_2)$ for any real θ ; but for a proper choice of θ , $x + 2e^{i\theta}\delta_2 \notin K(z_1,2\delta_1 + \delta_2)$, contradicting $K(z_2,5\delta_2) \subset K(z_1,2\delta_1 + \delta_2)$. Thus, $X_1{}^o \cap K(z_2,\delta_2) = X^o \cap K(z_2,\delta_2)$, so that $CX_1{}^o \cap K(z_2,\delta_2) = CX^o \cap K(z_2,\delta_2)$. We have, therefore, $\alpha(CX^o \cap K(z_2,\delta_2)) > 2\alpha(CX \cap K(z_2,5\delta_2))$. Finally, since $5\delta_2 \leq 2\delta_1 + \delta_2$, $\delta_2 \leq \delta_1/2$.

Proceeding in this manner we obtain sequences $\{z_n\}$, $\{r_n\}$, and $\{\delta_n\}$ such that $r_n \to \infty$, $K(z_{n+1},2\delta_{n+1}) \subset K^o(z_n,2\delta_n)$, $\delta_{n+1} \leq \delta_n/2$, and $\alpha(CX^o \cap K(z_n,\delta_n)) > n\alpha(CX \cap K(z_n,r_n\delta_n))$.

Clearly, $\delta_n \to 0$. Let $z_o = \bigcap K(z_n, 2\delta_n)$. Since $|z_n - z_o| < 2\delta_n$ we have $\alpha(CX^o \cap K(z_o, 3\delta_n)) \geq \alpha(CX^o \cap K(z_n, \delta_n))$ and $\alpha(CX \cap K(z_o, (r_n-2)\delta_n)) \leq \alpha(CX \cap K(z_n, r_n\delta_n))$. Therefore,

$$\frac{\alpha(CX^o \cap K(z_o, 3\delta_n))}{\alpha(CX \cap K(z_o, (r_n-2)\delta_n))} \geq \frac{\alpha(CX^o \cap K(z_n, \delta_n))}{\alpha(CX \cap K(z_n, r_n\delta_n))} \geq n .$$

Now let $r \geq 1$ be fixed and set $\varepsilon_n = 3\delta_n$. Since $r_n \to \infty$ as $n \to \infty$ we have, for large enough n,

$$\frac{\alpha(CX^o \cap K(z_o, \varepsilon_n))}{\alpha(CX \cap K(z_o, r\varepsilon_n))} \geq \frac{\alpha(CX^o \cap K(z_o, \varepsilon_n))}{\alpha(CX \cap K(z_o, (\frac{r_n-2}{3})\varepsilon_n))} \geq n ,$$

so that (4) fails to hold.

This completes the proof of 12.1.

The construction employed in the proof of (5) => (1) above can be used to prove other approximation theorems. For instance, we have

<u>Theorem</u> 12.2. Let $X \subset \mathbb{C}$ be compact, $f \in A(X)$. Then

f can be approximated uniformly on X by functions in $C(S^2)$ which are analytic on X^o and on a neighborhood of $\partial\Omega(X)$.

Proof. Exercise. (See 13.2).

Theorem 12.1 gives necessary and sufficient conditions on X that (the restriction to X of) every function in $C(S^2)$, analytic on X^o, belong to $R(X)$. One can pose a slightly different question and ask what conditions *on* *f* ($\in C(S^2)$) insure that f can be approximated uniformly on X by rational functions. The answer is contained in

Theorem 12.3. Let $X \subset \mathbb{C}$ be compact and $f \in C(S^2)$. Then $f \in R(X)$ if and only if there exists a constant $r \geq 1$ such that for all z and all $\delta > 0$

$$\left|\int_{\partial K(z,\delta)} f(\zeta)d\zeta\right| \leq \gamma(CX \cap K(z,r\delta))\Omega(\delta) ,$$

where $\Omega(\delta) \to 0$ as $\delta \to 0$.

Proof. [93] and [95].

13. APPLICATIONS OF VITUSHKIN'S THEOREM

As we remarked earlier, Vitushkin's theorem (12.1) contains as a special case virtually every other major theorem on the possibility of rational approximation on compact planar sets. It is now time to justify this remark and to make good our promise to prove the results stated in section 9.

The following lemma is basic.

Lemma 13.1. Let $X \subset \mathbb{C}$ be compact. Suppose $\underline{\lim}_{\delta\to 0} \alpha(CX \cap K(z,\delta))/\delta > 0$ for every $z \in \partial X$. Then $R(X) = A(X)$.

Proof. For any $r \geq 1$

$$0 < \underline{\lim}_{\delta\to 0} \frac{\alpha(CX \cap K(z,\delta))}{\delta} \leq \underline{\lim}_{\delta\to 0} \frac{\alpha(CX \cap K(z,\delta r))}{\alpha(CX^{o} \cap K(z,\delta))} .$$

Hence, condition (4) of 12.1 is satisfied for each point of ∂X. We are done.

Theorem 13.2. Let $X \subset \mathbb{C}$ be compact and suppose that each point of ∂X lies on the boundary of a component of CX. Then $R(X) = A(X)$.

Proof. Suppose the condition of the theorem is satisfied,

and let $z \in \partial X$. Then, for sufficiently small δ , $CX \cap K^o(z,\delta)$ contains an arc of diameter $\delta/2$. By 3.6 and 10.4 $\alpha(CX \cap K^o(z,\delta)) = \gamma(CX \cap K^o(z,\delta)) \geq \delta/8$. Thus the condition of 13.1 is satisfied.

Theorem 13.2 has as its corollaries the "standard" results of (qualitative) rational approximation theory. For instance, we have

<u>Theorem</u> 13.3. (Mergelyan) Let $X \subset \mathbb{C}$ be compact and suppose CX consists of a finite number of components. Then $R(X) = A(X)$.

<u>Corollary</u> 13.4. (Mergelyan) Let $X \subset \mathbb{C}$ be compact and suppose $\Omega(X) = S^2 \setminus X$, i.e., X does not divide the plane. Then every function in $A(X)$ is uniformly approximable on X by polynomials (in z).

<u>Proof</u>. This follows from 13.3 and a remark in section 9.

When 12.1 is specialized to sets without interior we obtain the following (earlier) result of Vitushkin [91].

<u>Theorem</u> 13.5. Let $X \subset \mathbb{C}$ be compact. The following are equivalent:

(1) $R(X) = C(X)$.

(2) $\gamma(G \setminus X) = \gamma(G)$, for any bounded open set G .

(3) $\gamma(K^{o}(z,\delta) \setminus X) = \gamma(K^{o}(z,\delta)) = \delta$ for all z and δ .

Proof. This follows immediately from 12.1 We need only note that $R(X) = C(X)$ implies that $X^{o} = \emptyset$ and recall that $\gamma(U) = \alpha(U)$ for open sets (10.3).

Actually, one can prove $R(X) = C(X)$ under conditions even weaker than (3) above. For instance, 13.1 gives a better sufficient condition than (2) or (3). An even weaker sufficient condition for $R(X)$ to coincide with $C(X)$ is that

$$\overline{\lim_{\delta\to 0}} \frac{\gamma(K^{o}(z,\delta)\setminus X)}{\delta^{2}} = \infty$$

for almost all (dxdy) $z \in X$. We shall not prove this statement here; a reference is [41].

The theorem of Hartogs and Rosenthal is an easy consequence of 13.5.

Theorem 13.6. (Hartogs-Rosenthal). Let $X \subset \mathbb{C}$ be compact and suppose X has zero Lebesgue planar measure. Then $R(X) = C(X)$.

Proof. Let D_{δ} be an open disc of radius δ . We must show that $\gamma(D_{\delta} \setminus X) = \delta$. Clearly, $\gamma(D_{\delta} \setminus X) \leq \delta$. Let K_{ε} be a closed subset of $D_{\delta} \setminus X$ such that $V(K_{\varepsilon}) \geq \pi\delta^{2}-\pi\varepsilon$

$(0 < \varepsilon < \delta^2)$; here $V(K_\varepsilon)$ is the area (=Lebesgue planar measure) of K_ε . By 3.12, $V(K_\varepsilon) \leq \pi[\gamma(K_\varepsilon)]^2$. Thus $\sqrt{\delta^2-\varepsilon} \leq \gamma(K_\varepsilon) \leq \gamma(D_\delta \setminus X)$. Letting $\varepsilon \to 0$, we obtain $\delta \leq \gamma(D_\delta \setminus X)$, as required.

Exercise. Prove that

$$\overline{\lim_{\delta \to 0}} \frac{\gamma(K^o(z,\delta)\setminus X)}{\delta} > 0$$

for almost all $z \in X$ implies that $R(X) = C(X)$.

14. GEOMETRIC CONDITIONS

Because the situation with which it deals admits great topological complications, the statement of Vitushkin's theorem was framed necessarily in terms of the nonituitive, quasi-geometrical notion of AC capacity. It is natural to expect that, in particularly nice cases, the actual geometry of X will play a significant role in determining whether $R(X) = A(X)$. This is indicated by the examples at the end of section 9 as well as by 13.2. In this section, we shall discuss what is known in this direction.

Let X be compact and let $CX = \bigcup_{i=0}^{\infty} U_i$ be the decomposition of CX into (open) components. We call

$$\Gamma_I = \partial X \setminus \bigcup_{i=0}^{\infty} \partial U_i$$

the <u>inner boundary</u> of X. In [94], Vitushkin noted that if Γ_I consists of a finite number of points then $R(X) = A(X)$ This shows, in particular, that if X is a set of type (L) (section 8) then $R(X) = A(X)$. Actually, much more is true. For instance, using the results of sections 5 and 12 it is not hard to prove

<u>Theorem</u> 14.1. (Melnikov [64]). Suppose Γ_I is a subset

of an analytic curve. Then $R(X) = A(X)$.

Proof. We shall verify condition (4) of 12.1 with $r = 1$ and a uniform bound m depending on the analytic curve in question. For $z \notin \Gamma_I$ it is easy to see that this condition is satisfied, so it is enough to prove

$$(1) \qquad \alpha(CX^o \cap K(z,\delta)) \leq m\alpha(CX \cap K(z,\delta))$$

for $z \in \Gamma_I$ and δ sufficiently small. Accordingly, fix $z \in \Gamma_I$ and let $K = K(z,\delta)$, where δ is small. Since the condition (1) is local, we may assume (employing an appropriate conformal map, if necessary) that Γ_I lies on a circle σ . σ divides the plane into two components C_1 (the bounded component) and C_2 . Let $T_j = \overline{C}_j$, $j = 1,2$ and set $\Gamma_j = \partial(T_j \cap K)$.

Now choose $\varphi \in C(CX^o \cap K,1)$ such that $\varphi'(\infty) = (1/2)\alpha(CX^o \cap K)$. Clearly,

$$(2) \qquad \int_{\partial K} \varphi(\zeta)d\zeta = \int_{\Gamma_1} \varphi(\zeta)d\zeta + \int_{\Gamma_2} \varphi(\zeta)d\zeta \, .$$

By 12.2 , φ can be approximated uniformly on T_1 by functions analytic on $T_1 \setminus E_1$, where E_1 is a closed subset of $CX \cap K$ lying in $T_1{}^o = C_1$. Moreover, the approximating functions can be taken to satisfy $\|f\|_{T_1} \leq 1$. For such functions f we have, from 10.6 ,

$$|\int_{\Gamma_1} f(\zeta)d\zeta| = |\int_\sigma f(\zeta)d\zeta| \leq c\alpha(E_1) \leq c\alpha(CX \cap K) ,$$

where c is a universal constant. Hence

$$(3) \qquad |\int_{\Gamma_1} \varphi(\zeta)d\zeta| \leq c\alpha(CX \cap K) .$$

Similarly,

$$(4) \qquad |\int_{\Gamma_2} \varphi(\zeta)d\zeta| \leq c\alpha(CX \cap K) .$$

It follows from (2), (3), and (4) that

$$\alpha(CX^o \cap K) = \pi^{-1}|\int_{\partial K} \varphi(\zeta)d\zeta| < c\alpha(CX \cap K) ,$$

as required.

Exercise. The preceding proof contains a suppressed compactness argument. Supply it .

Vitushkin [95] has improved 14.1. To state his theorem we need the following

Definition 14.2. Let σ be a curve without self-intersections and suppose that σ can be decomposed into a finite number of Jordan curves $\sigma_1,\ldots,\sigma_n$ such that for each i σ_i has a well-defined tangent vector $\tau(z,\sigma_i)$ at every point $z \in \sigma_i$. Suppose further that for each i the function $\arg \tau(z,\sigma_i)$ obeys a Hölder condition with constant L and exponent $\alpha > 0$:

$$|\arg \tau(z,\sigma_i) - \arg \tau(z',\sigma_i)| \leq L|z-z'|^{\alpha} \qquad z,z' \in \sigma_i .$$

Then we call σ a <u>Liapunov curve</u>.

Vitushkin's extension of 14.1 can now be formulated as

<u>Theorem</u> 14.3. Let Γ_I lie on the union of countably many Liapunov curves. Then $R(X) = A(X)$.

We shall not prove 14.3, since the proof is long and complicated. What one must establish is the analogue of 5.11 for Liapunov curves. Once this has been accomplished, 14.3 follows from 12.1

The method of 9.6 shows that if $\partial X \backslash \Gamma_I$ has finite length and $R(X) = A(X)$ then $\alpha(\Gamma_I) = 0$. This is no longer true if the length of $\partial X \backslash \Gamma_I$ is infinite. Indeed, we have

<u>Example</u> 14.4. Let $J \subset \Delta$ be an arc of positive measure (cf. 9.10). Let U_i, $i = 1,2,\ldots$, be a sequence of open sets such that

(1) $\overline{U}_i \subset \Delta \setminus J$ for all i;

(2) $\overline{U}_i \cap \overline{U}_j = \emptyset$ unless $i = j$;

(3) diameter $(U_i) > \varepsilon > 0$ for some ε independent of i ;

(4) J is the set of "limit points" of the set $\bigcup_{i=1}^{\infty} \partial U_i$.

Let $X = \overline{\Delta} \setminus \bigcup_{i=1}^{\infty} U_i$. By (1) and (4) , $\Gamma_I = J$ so that $\alpha(\Gamma_I) > 0$. However, it follows from (3) and 15.10 that $R(X) = A(X)$. (It is obvious from (3) that $\partial X \setminus \Gamma_I$ does not have finite length. On the other hand, of course, each ∂U_i may be rectifiable.)

The existence of the phenomenon exhibited by 14.4 was first pointed out to me by Alfred Hallstrom .

15. FUNCTION ALGEBRA METHODS

In the preceding sections we met with considerable success in applying constructive methods to problems of rational approximation. No discussion of this subject, however, would be complete without an indication of how the (nonconstructive) methods of functional analysis can be employed to study rational approximation. Accordingly, this section is devoted to the development and exploitation of this approach. At the same time, we can prove results of some interest in themselves. In particular, we offer a proof of Bishop's localization theorem (15.9); we also give a proof, due to Garnett, of an interesting special case of 14.1.

Let X be a compact set in the plane. Suppose $f \in C(X)$ and μ is a complex Baire measure on X. If μ kills f, i.e., if $\int_X f\, d\mu = 0$, we write $\mu \perp f$. If $\mu \perp f$ for every $R(X)$ (resp. $A(X)$), we say μ <u>annihilates</u> $R(X)$ (resp. $A(X)$) and call μ an <u>annihilating measure</u> for $R(X)$ (resp. $A(X)$). In this case we write $\mu \perp R(X)$ (resp. $\mu \perp A(X)$). Now let $f \in C(X)$ be fixed. Since $R(X)$ is a closed subspace of the Banach space $C(X)$, it follows (cf. section 2) from standard theorems of functional analysis that $f \in R(X)$ if and only if $\mu \perp f$ for every measure μ which annihilates $R(X)$. Thus $R(X) = A(X)$

if and only if they have the same annihilating measures. In particular, $R(X) = C(X)$ if and only if the only measure which annihilates $R(X)$ is 0.

Following Bishop [13], we shall study annihilating measures for $R(X)$ by examining their "Cauchy transforms." Let μ be a finite complex Baire measure with compact support K. Set

$$\hat{\mu}(z) = \int (\zeta - z)^{-1}\, d\mu(\zeta) .$$

Then $\hat{\mu}$ is clearly analytic off K; moreover, an interchange of order of integration shows that $\hat{\mu}$ is locally integrable with respect to Lebesgue planar measure. This gives

<u>Lemma</u> 15.1. Let μ be a positive measure on X and let

$$N(z) = \int \frac{d\mu(\zeta)}{|\zeta - z|} .$$

Then $N(z) < \infty$ almost everywhere $(dxdy)$.

<u>Lemma</u> 15.2. $\hat{\mu} = 0$ almost everywhere if and only if $\mu = 0$.

<u>Proof</u> [16]. One direction is trivial. Let g be a continuously differentiable function which vanishes outside a compact set containing the support K of μ. By Green's theorem

$$g(z) = -\frac{1}{\pi}\iint \frac{\partial g}{\partial \bar{\zeta}} \frac{1}{\zeta - z}\, dxdy ,$$

where $\zeta = x + iy$ and the integration is extended over the whole complex plane (or any large compact set). Interchanging orders of integration, we obtain

$$\int_K g(z)d\mu(z) = -\frac{1}{\pi}\iint \frac{\partial g}{\partial \bar{\zeta}} \left\{\int_K \frac{d\mu(z)}{\zeta - z}\right\} dxdy$$

$$= \frac{1}{\pi}\iint \frac{\partial g}{\partial \bar{\zeta}} \hat{\mu}(\zeta)\, dxdy$$

$$= 0$$

since $\hat{\mu} = 0$ almost everywhere. Since any function continuous on K can be approximated uniformly on K by such functions g, μ kills every function in $C(K)$. By the Riesz representation theorem, $\mu = 0$.

Lemma 15.3. Let $X \subset \mathbb{C}$ be compact and let μ be a measure supported on X. A necessary and sufficient condition that $\mu \perp R(X)$ is that $\hat{\mu} = 0$ on $\mathbb{C} \setminus X$.

Proof. Necessity is clear. For sufficiency, suppose $\hat{\mu} = 0$ on $\mathbb{C} \setminus X$. Then μ kills $(\zeta - z)^{-1}$ for each $z \in \mathbb{C} \setminus X$. It follows from a partial fraction decomposition that μ kills any rational function with simple poles, all of which lie off X. Since any element of $R(X)$ can be approximated by such functions, the lemma follows.

Lemma 15.3 can be used to give a neat proof of the following extension of Runge's theorem (9.1).

Theorem 15.4. Suppose $f \in A(X)$ and f can be extended to be continuously differentiable in a neighborhood of X. Then $f \in R(X)$.

Proof. We may suppose f to be a continuously differentiable function of compact support in the plane. Then

$$f(z) = -\frac{1}{\pi}\iint \frac{\partial f}{\partial \bar\zeta}\frac{1}{\zeta - z}\,dxdy \qquad (\zeta = x+iy),$$

where the integral is taken over the whole plane. Suppose $\mu \perp R(X)$; then $\hat\mu = 0$ on $\mathbb{C} \setminus X$ by 15.3. Since f is analytic on X^o, $\frac{\partial f}{\partial \bar\zeta} = 0$ there; hence, by continuity, $\frac{\partial f}{\partial \bar\zeta} = 0$ on X. Thus $\hat\mu \cdot \frac{\partial f}{\partial \bar\zeta} = 0$ everywhere. Therefore,

$$\int_X f(z)d\mu(z) = \frac{1}{\pi}\iint \hat{u}(\zeta)\,\frac{\partial f}{\partial \bar\zeta}\,dxdy = 0 .$$

The above proof is due to Andrew Browder (unpublished).

Using 15.1 and 15.2 we can prove 2.4.

Theorem 15.5. (Bishop [13]). If almost every (dxdy) point of X is a peak point for $R(X)$ then $R(X) = C(X)$.

Proof [100]. Let $\mu \perp R(X)$ and suppose $\mu \neq 0$. Denote by P the set of peak points of $R(X)$. Let

$$N(z) = \int \frac{d|\mu|(\zeta)}{|\zeta - z|} .$$

Then $N(z) < \infty$ almost everywhere dxdy. Since $\hat{\mu} = 0$ on $\mathbb{C} \setminus X$ (15.3) and $\mu \neq 0$ it follows that $\hat{\mu} \neq 0$ on a subset of X having positive measure. Now almost every point of X belongs to P ; hence we can choose $z_0 \in P$ such that $N(z_0) < \infty$ and $\hat{\mu}(z_0) \neq 0$. From $N(z_0) < \infty$ it follows that $|\mu|(\{z_0\}) = 0$. Now if r is a rational function with poles off X so is $[r(z)-r(z_0)]/(z-z_0)$; hence

$$\int_X \frac{r(z)-r(z_0)}{z-z_0} \, d\mu(z) = 0$$

or

$$\int_X \frac{r(z)}{z-z_0} \, d\mu(z) = r(z_0)\hat{\mu}(z_0) ,$$

where both sides of the above equation converge absolutely

by the choice of z_0. It follows that

$$\int \frac{f(z)}{z-z_0}\,d\mu(z) = f(z_0)\hat{\mu}(z_0)$$

for all $f \in R(X)$. Let g be a function in $R(X)$ which peaks at z_0. Then

$$(*) \qquad \int \frac{g^k(z)}{z-z_0}\,d\mu(z) = \hat{\mu}(z_0)$$

for every positive integer k. Since $\mu(\{z_0\}) = 0$, the integrand on the l.h.s. tends to 0 almost everywhere $d\mu$ as $k \to \infty$. Also,

$$\left|\frac{g^k(z)}{z-z_0}\right| \leq \frac{1}{|z-z_0|},$$

where the r.h.s. belongs to $L^1(d|\mu|)$ by the choice of z_0. Letting $k \to \infty$ in (*) and applying the dominated convergence theorem, we get

$$0 = \hat{\mu}(z_0),$$

a contradiction. Hence $\mu = 0$ and $R(X) = C(X)$.

An immediate corollary is

Theorem 15.6. $R(X) = C(X)$ if and only if every point of X is a peak point for $R(X)$.

Recently, Wilken [107] proved a nice generalization of this result (see the bibliographical notes). It is worth observing that 15.5 combined with Melnikov's theorem (6.1) yields the following

Theorem 15.7. $R(X) = C(X)$ if and only if for almost every $x \in X$

$$\sum_{n=0}^{\infty} 2^n \gamma_n(x) = \infty ,$$

where $\gamma_n(x) = \gamma(CX \cap \{2^{-n-1} \leq |z-x| \leq 2^{-n}\})$.

This improves a result of Vitushkin [89], who proved that if $\sum 4^n \gamma_n(x) = \infty$ for every $x \in X$ then $R(X) = C(X)$. Finally, let us note that the theorem of Hartogs and Rosenthal (13.6) is an immediate corollary of 15.5.

In 11.8 we proved the remarkable result that questions of rational approximation are "local." Using the ideas of this section we can offer another proof of this fact. First, however, we need

Lemma 15.8. Let φ be a C^{∞} function with compact support K and let μ be a measure supported on X. Then

$$\varphi\hat{\mu} = \widehat{\varphi\mu} + \hat{\sigma} ,$$

where

$$\sigma = -\frac{1}{\pi}\frac{\partial\varphi}{\partial\bar{\zeta}}\,\hat{\mu}(\zeta)\;dxdy .$$

Proof. Interchanging order of integration and applying Green's theorem, we have

$$\hat{\sigma}(z) = -\frac{1}{\pi}\iint_K \frac{\partial\varphi}{\partial\bar{\zeta}} \{\int_X \frac{d\mu(t)}{t-\zeta}\} \frac{1}{\zeta-z}\,dxdy$$

$$= \int_X \{-\frac{1}{\pi}\iint_K \frac{\partial\varphi}{\partial\bar{\zeta}} \frac{1}{t-\zeta}\frac{1}{\zeta-z}\,dxdy\}\,d\mu(t)$$

$$= \int_X \frac{1}{t-z}\{-\frac{1}{\pi}\iint_K \frac{\partial\varphi}{\partial\bar{\zeta}} [\frac{1}{t-\zeta} + \frac{1}{\zeta-z}]\,dxdy\}\,d\mu(t)$$

$$= \int_X \frac{1}{t-z}[-\varphi(t) + \varphi(z)]\,d\mu(t)$$

$$= -\widehat{\varphi\mu}(z) + \varphi(z)\hat{\mu}(z) .$$

Theorem 15.9. (Bishop) Let $X \subset \mathbb{C}$ be compact and let $f \in C(X)$. Suppose each point $x \in X$ has a closed neighborhood U_x (in X) such that the restricted function $f|_{U_x}$ belongs to $R(U_x)$. Then $f \in R(X)$.

Proof. Since X is compact we can find $x_1, \ldots, x_n$ in X such that the sets $U_j = U_{x_j}$ form a covering of X. Suppose $\mu \perp R(X)$. Choose a (finite C^∞) partition of unity $\{\varphi_j\}_{j=1}^n$ such that φ_j has compact support and $\varphi_j \equiv 0$ on $X \setminus U_j$. Let

$$\nu_j = \varphi_j \mu - \frac{1}{\pi} \frac{\partial \varphi_j}{\partial \bar{\zeta}} \hat{\mu}(\zeta)\, dxdy \qquad (\zeta = x+iy)$$

Since μ is supported on X, it follows from 15.3 that ν_j is supported on U_j. Moreover, by 15.8, $\hat{\nu}_j = \varphi_j \hat{\mu}$, so that $\hat{\nu}_j = 0$ off U_j. By 15.3, $\nu_j \perp R(U_j)$. Therefore, $\nu_j \perp f$ $j = 1,2,\ldots,n$. Since $\mu = \sum_{j=1}^n \nu_j$, we obtain $\mu \perp f$, as required.

Bishop never published his proof of 15.9, the heart of which is lemma 6 of [10]. The proof given above is an elaboration of the one in [34], which is there attributed to Hoffman. Recently, Garnett [34] observed that 15.9 can be used to prove the following corollary of 12.1, first stated in [67].

Theorem 15.10. Let $X \subset \mathbb{C}$ be compact. Suppose the diameters of the components of CX are bounded away from 0. Then $R(X) = A(X)$.

Proof. Let $\varepsilon > 0$ be a lower bound for the diameters of the components of CX. For each $x \in X$ choose a closed neighborhood U_x (in X) whose diameter is less than ε. Clearly $S^2 \setminus U_x$ is connected; hence, by 13.4, $A(U_x) = R(U_x)$. Thus, if $f \in A(X)$ we have $f|_{U_x} \in R(U_x)$. It follows from 15.9 that $f \in R(X)$.

Mergelyan's theorem on rational approximation (13.3) is an immediate consequence of 15.10. Note, however, that in the proof of 15.10 we needed to use the polynomial approximation theorem (13.4), while in the development of section 13 13.4 was obtained as a special case of 13.3. A proof of 13.4 in the spirit of the present section is in [16].

Quite recently, Garnett showed that the methods of this section can be used to prove a theorem of the kind discussed in section 14. His result generalizes a theorem of Vitushkin and is included in 13.2 and 14.1.

Theorem 15.11. Let $X \subset \mathbb{C}$ be compact. Let Γ_0 be the set of all points in X each neighborhood of which intersects infinitely many components of CX. Then if Γ_0 is a countable set $A(X) = R(X)$.

Proof. First, suppose Γ_0 is a singleton, say 0. Let φ be a C^∞ function such that (a) $\varphi(z) = 1$, if $|z| \leq 1/2$;

(b) $\varphi(z) = 0$, if $|z| \geq 1$; and (c) $\| \frac{\partial \varphi}{\partial \bar{z}} \|_\infty \leq K$. Let $\varphi_n(z) = \varphi(2^n z)$. Then (a') $\varphi_n(z) = 1$, if $|z| \leq 2^{-n-1}$; (b') $\varphi_n(z) = 0$, if $|z| \geq 2^{-n}$; and (c') $\| \frac{\partial \varphi_n}{\partial \bar{z}} \|_\infty \leq K2^n$.

By Green's theorem

$$\varphi_n(z) = -\frac{1}{\pi} \int_{A_n} \frac{\partial \varphi_n}{\partial \bar{\zeta}} \frac{1}{\zeta - z} \, dxdy ,$$

where $\zeta = x+iy$ and $A_n = \{2^{-n-1} \leq |\zeta| \leq 2^{-n}\}$. Now suppose $\mu \perp R(X)$. Set

$$\nu_n = \varphi_n \mu - \frac{1}{\pi} \frac{\partial \varphi_n}{\partial \bar{\zeta}} \hat{\mu}(\zeta) \, dxdy .$$

Then ν_n is supported on A_n and by 15.8 we have $\hat{\nu}_n = \varphi_n \hat{\mu}$. Let $X_n = X \cap \{|z| \geq 2^{-n-1}\}$. By 15.3, $\widehat{\mu - \nu_n} = (1-\varphi_n)\hat{\mu} = 0$ on CX_n . Applying the other half of 15.3, we obtain $\mu - \nu_n \perp R(X_n)$. Since CX_n consists of only finitely many components $R(X_n) = A(X_n)$, by 15.10. Hence $\mu - \nu_n \perp A(X_n)$, and so $\mu - \nu_n \perp A(X)$.

Suppose $f \in A(X)$; we wish to show $\mu \perp f$. Since $\mu = (\mu - \nu_n) + \nu_n$ and $\mu - \nu_n \perp A(X)$, we have

$$\int_X f \, d\mu = \int_X f \, d\nu_n \tag{1}$$

for all n . Since $\nu_n \perp R(X)$, $\nu_n \perp 1$; therefore,

$$\int_X [f-f(0)]\, d\nu_n = \int_X f\, d\nu_n \tag{2}$$

for any $f \in C(X)$. An easy computation shows that the sequence $\{\|\nu_n\|\}$ is bounded. Therefore,

$$\int_X [f-f(0)]\, d\nu_n \leq \| f-f(0)\|_{A_n} \|\nu_n\| \to 0 \tag{3}$$

as $n \to \infty$, since f is continuous. It follows from (1), (2), and (3) that

$$\int_X f\, d\mu = 0 ,$$

as required.

Now suppose Γ_0 is countable and let $f \in A(X)$ be given. Let S be the set of points of Γ_0 at which f is not locally in $R(X)$. Suppose $S \neq \emptyset$. Then since S is countable and compact, it has an isolated point y . But the argument above shows that f must belong to $R(X)$ locally at y , a contradiction. Therefore, $S = \emptyset$. By 15.9, we are done.

<u>Exercise</u>. Give a constructive proof of 15.11.

There is an alternate formulation of 15.11 that is worth mentioning. Say $x \in N(X)$, the <u>nucleus</u> of X , if there exists no closed neighborhood K of x (in X) for which $R(K) = A(K)$. Clearly, $N(X)$ is a closed subset of Γ_0 . The argument of 15.11 then shows that $N(X)$ must be a perfect set.

In conclusion, let us note that function algebra ideas considerable more elaborate than the simple techniques exploited in this section have also proved successful in studying rational approximation. The interested reader should consult [103] for an exposition of these methods. A more extended treatment will appear in the forthcoming book <u>Uniform Algebras</u>, by T. W. Gamelin.

16. SOME OPEN QUESTIONS

To the best of my knowledge, the following problems remain open.

16.1. Let X be compact, $x \in X$. Let I denote the set of numbers $t \in [0,1]$ for which $\{|z-x| = t\} \cap CX \neq \emptyset$.

<u>Conjecture</u> If $\int_I t^{-1}\,dt = \infty$, x is a peak point for $R(X)$.

This is true for sets of type (L). If the conjecture is valid in general it sharpens (via the theorem of Keldysh) a sufficient condition for regularity for the Dirichlet problem due to Beurling.

It is easy to see that the condition in the conjecture is not necessary for x to be a peak point.

16.2. <u>Conjecture</u>. There exists an absolute constant c such that $\gamma(S_1 \cup S_2) \leq c[\gamma(S_1) + \gamma(S_2)]$ for all sets S_1, S_2. (Cf. 5.9 and 5.12).

This may well be false, but it would be good to have a proof or a counterexample.

16.3. <u>Conjecture</u>. There exists an absolute constant c such that $\alpha(S_1 \cup S_2) \leq c[\alpha(S_1) + \alpha(S_2)]$ for all sets S_1, S_2.

16.4. Let X be compact, $x \in X$.

<u>Conjecture</u>. x is a peak point for $A(X)$ if and only if $\sum_{n=0}^{\infty} 2^n \alpha_n = \infty$, where $\alpha_n = \alpha(CX^o \cap \{2^{-n-1} \leq |z-x| \leq 2^{-n}\})$.

P. C. Curtis has recently proved the sufficiency of the above condition [114].

16.5. <u>Conjecture</u>. $A(X) = R(X)$ if and only if they have the same peak points.

We have noted (9.8) that this cannot be weakened to the requirement that the peak points coincide except for a set of measure zero.

16.6. Let X be compact and let $\{U_i\}$ be the set of components of CX. Recall that the inner boundary Γ_I of X is $\Gamma_I = \partial X \setminus (\bigcup_{i=0}^{\infty} \partial U_i)$.

<u>Conjecture</u>. If $\alpha(\Gamma_I) = 0$, then $A(X) = R(X)$.

We have discussed some special cases of this conjecture in section 14, where we also noted that the converse is false. If the conjecture of 16.3 is correct, then the conjecture of this paragraph is also true ([94]).

APPENDIX I. LOGARITHMIC CAPACITY

In this appendix we bring together for the reader's convenience several (equivalent) definitions of logarithmic capacity; we also justify a statement in 3.5. A detailed treatment of the material of this section is in [84].

Let K be a compact set in the plane. Let μ be a (positive Baire) probability measure supported on K and set

$$I(\mu) = \iint \log\left|\frac{1}{z-\zeta}\right| d\mu(\zeta)d\mu(z)$$

Let $V = \inf I(\mu)$, where the inf is taken over all such measures. Then $-\infty < V \leq \infty$. The logarithmic capacity of K is given by

$$\text{cap}\,(K) = e^{-V},$$

where e^{-V} is understood to be 0 if $V = +\infty$. (This definition, now standard, is somewhat different from that used by early writers in the subject. Accordingly, the reader should exercise some caution in reading the older literature.) As usual, one extends cap to arbitrary sets by

$$\text{cap}(S) = \sup_{K \subset S} \text{cap}(K).$$

A second way of defining $\text{cap}(K)$ is to set

$$U_\mu(z) = \int \log\left|\frac{1}{z-\zeta}\right| d\mu(\zeta) ,$$

where μ is a Baire probability measure on K. Then

$$V = \inf_{\mu} \sup_{\Omega(K)} U_\mu(z) ,$$

where the inf is taken over all such measures. As before, we have $\operatorname{cap}(K) = e^{-V}$.

If K is a compact set and $z_1, z_2, \ldots, z_n \in K$ let

$$V(z_1, \ldots, z_n) = \prod_{j<k} |z_j - z_k| .$$

Set $V_n = \max V(z_1, \ldots, z_n)$, where the maximum is taken over all n-tuples of points in K. Then if $d_n = V_n^{1/\binom{n}{2}}$, d_n is a (weakly) decreasing sequence with limit $\tau(K)$. We call $\tau(K)$ the <u>transfinite</u> <u>diameter</u> of K. A beautiful theorem of Szegö [84] tells us that $\tau(K) = \operatorname{cap}(K)$. This identity is often useful in estimating the capacities of sets.

There are still other methods by which the logarithmic capacity of a compact set can be defined. For instance, one definition hinges on the use of Tchebycheff polynomials. These definitions need not concern us here. We will, however, need the following fact, a proof of which is in [84]. Let $g(z,\infty)$ be the Green's function for $\Omega(K)$ with a pole at ∞;

then

$$\lim_{z\to\infty} (\log|z|-g(z,\infty)) = \log \operatorname{cap}(K)$$

Proposition I.1. Let K be compact. Then

$$\operatorname{cap}(K) = \sup|f'(\infty)| ,$$

where the sup is taken over all functions analytic (but not necessarily single-valued) on $\Omega(K)$ which satisfy

(1) $|f|$ is single-valued;

(2) $|f(z)| < 1$, $z \in \Omega(K)$;

(3) $f(\infty) = 0$.

Proof. If the sup is zero, $\Omega(K)$ supports no bounded harmonic function. Therefore $\operatorname{cap}(K) = 0$. For otherwise there is a positive measure μ such that

$$u(z) = \int \log\left|\frac{1}{z-\zeta}\right| d\mu(\zeta)$$

is a bounded harmonic function, contrary to hypothesis. Suppose then that f satisfies (1) through (3) above and $|f'(\infty)| > 0$. We can assume that $0 \notin \Omega(K)$. Let $\{z_n\}$ be the set of finite zeros of f . Since f is analytic at ∞ the z_n can accumulate only at $\partial\Omega(K)$. Let $\{D_n\}$

be a sequence of domains such that $D_1 \subset D_2 \subset \ldots \to \Omega(K)$ and ∂D_n consists of a finite number of (disjoint) simple closed analytic curves. Let $g_n(z,\infty)$ be the Green's function for D_n with pole at ∞. Then $g_n(z,\infty)-\log|z| \to g(z,\infty)-\log|z|$ uniformly on subsets of $\Omega(K)$ compact in the topology of the sphere, as $n \to \infty$. Now let $u(z) = -\log|f(z)|$. $u(z)$ is a positive superharmonic function on $\Omega(K)$. Since $u(z)-\log|z|$ is harmonic near ∞ it also is superharmonic on $\Omega(K)$. Now $u(z) \geq g_n(z,\infty)$ on ∂D_n since u is positive and $g_n = 0$ on ∂D_n. Therefore, $u(z)-\log|z| \geq g_n(z,\infty)-\log|z|$ on ∂D_n. The l. h. s. is a superharmonic function while the r. h. s. is harmonic on D_n. Thus $u(z)-\log|z| \geq g_n(z,\infty)-\log|z|$ on D_n. Letting $n \to \infty$ we obtain

$$u(z)-\log|z| \geq g(z,\infty)-\log|z| \qquad z \in \Omega(K)$$

Thus

$$|f'(\infty)| = \lim_{z\to\infty}|zf(z)| = \lim_{z\to\infty}|ze^{-u(z)}|$$

$$= \lim_{z\to\infty} \exp[\log|z|-u(z)]$$

$$\leq \lim_{z\to\infty} \exp[\log|z|-g(z,\infty)] = \operatorname{cap}(K)$$

by the remark preceding the proposition. If we set $\varphi(z) = \exp[-g(z,\infty)-ih(z)]$, where $h(z)$ is a (multiple-valued) harmonic conjugate of $g(z,\infty)$, it is immediate that $|\varphi'(\infty)| = \text{cap}(K)$. This completes the proof.

The importance of logarithmic capacity in function theory stems from the fact that sets of logarithmic capacity zero are "negligible sets" for harmonic functions. In particular, suppose U is an open set , $K \subset U$ is compact, and $\text{cap}(K) = 0$. Then any function bounded and harmonic on $U \setminus K$ can be extended harmonically to all of U .

APPENDIX II. ANALYTIC CAPACITY AND THE REMOVABILITY OF SINGULARITIES

This brief section is devoted to pointing out the significance of analytic capacity in problems of function theory.

A compact set K is said to be a Painlevé null set if $\Omega(K)$ supports no nonconstant bounded analytic function. Thus, Painlevé null sets are totally disconnected, since if $\sigma \subset K$ is a continuum the Riemann map of $S^2 \setminus \sigma$ onto Δ provides a nonconstant bounded analytic function on $\Omega(K)$.

<u>Proposition</u> II.1. The Painlevé null sets coincide with the compact sets of analytic capacity zero.

<u>Proof</u>. Clearly, if K is a Painlevé null set $\gamma(K) = 0$. On the other hand, let f be a bounded analytic function on $\Omega(K)$. We can choose a constant C and a nonnegative integer m such that $g(z) = Cz^m f(z)$ is an admissible function for K and $g'(\infty) \neq 0$. Thus $\gamma(K) \neq 0$.

Actually, something stronger is true: sets of analytic capacity zero are removable sets for bounded analytic functions. Although this follows from the remarks after 5.7, it is easy to give a simple independent proof.

Proposition II.2. Let K be a compact set and U an open set such that $K \subset U$. Then every function bounded and analytic on $U \setminus K$ can be extended analytically to all of U if and only if $\gamma(K) = 0$.

Proof. If $\gamma(K) \neq 0$ there exists a nonconstant function bounded and analytic on $S^2 \setminus K = \Omega(K)$. If this function were extendable to a function analytic on all of U it would be constant by Liouville's theorem. In the other direction, suppose $\gamma(K) = 0$. Let f be bounded and analytic on $U \setminus K$. Using the Cauchy integral formula we can write $f = f_1 + f_2$ where f_1 is analytic on all of U and f_2 is analytic on $S^2 \setminus K$. Since $\gamma(K) = 0$, $f_2 \equiv 0$.

In conclusion, let us note that Havinson [49] has used analytic capacity quite successfully to study other problems in function theory. The interested reader should consult his paper for details.

BIBLIOGRAPHICAL NOTES

A bibliography is a product of an author's whimsy and of his ignorance; the one that follows is no exception. That the list of references is complete in any strong sense is probably too much to hope for; however, to the best of my knowledge, all the basic references have been included, as well as a large number of papers that are, perhaps, of more tangential interest. I have tried especially to be generous in citations of the Soviet literature. Also, with the hope of stimulating some interest in the fascinating history of the subject, I have included a number of references to the early literature. These references, however, have been chosen rather subjectively and are not meant to be complete.

A word or two is in order at to the actual organization of the bibliography. The entries are grouped alphabetically by author. The transliteration of Russian names has been based on the transliterations in Doklady. If an article has received notice in Mathematical Reviews, this reference is given at the end of the corresponding entry. In general, I have not listed announcements of results when complete proofs have become available in subsequent papers; exceptions to this rule are cases in which the announcements are significantly more accessible. When a Russian paper has been translated into English I have referenced the translated paper.

Included here are also some brief notes, grouped under several headings, concerning the content, historical importance, or interrelationships of the papers referenced. These are by no means meant to be complete; their purpose is to whet, not satisfy, the reader's appetite for more.

1. Earlier work on rational approximation. Runge's original paper [78] marks the beginning of the systematic study of rational approximation. That same year, 1885, Weierstrass proved his famous theorem on polynomial approximation. Earlier, Appell [5],[6] had studied some special cases of approximability by rational functions. Basic progress in studying polynomial and rational approximation was made by Walsh [96],[97], Hartogs and Rosenthal [44] (cf. Tonyan [83]), Lavrentiev [60] (cf. Mergelyan [65]), and Keldysh [58]. We have commented on the contributions of these mathematicians in Section 9; for another discussion of this work, see the treatise of Mergelyan [67]. One should also consult the monograph of Walsh [99], which contains other references to the early literature.

2. More recent (Russian) work on rational approximation. The definitive results on polynomial and rational approximation are due to Mergelyan [67] and Vitushkin [93],[94],[95]. Mergelyan first proved his beautiful theorem on polynomial approximation in [66]; [67] contains a very full treatment and generalizations to rational approximation. Vitushkin's papers [88],[89],[91], [93],[94],[95] constitute a triumphant march toward the solution of the rational approximation problem; there are many misprints and incomplete proofs in these papers, however. [68] and [41] are useful summaries of the state of the subject in 1961 and 1965 respectively. The reader can also consult the book of Smirnov and Lebedev [81], which contains a nice exposition of

Mergelyan's results and the earlier work of Vitushkin. A similar treatment occurs in the difficult-to-obtain notes of Gamelin [31]. Other papers that here deserve mention are those of Dolzhenko [27],[28] and Gonchar [40].

3. Analytic capacity. The study of sets of analytic capacity zero goes back all the way to Painlevé [71] (see Appendix II). Refer also to the papers of Denjoy [22] and Besicovitch [8]. The fascinating question of priority in these matters is dealt with in [20].

The actual definition of analytic capacity is due to Ahlfors [2], who was interested in function theoretic extremal problems on finitely connected planar domains. His work was refined by Garabedian [33]. A summary of this and related work is in Nehari's survey article [70]. Ahlfors generalized his results to regions on Riemann surfaces [3]; see Royden's paper [77] for another treatment as well as further references to the literature.

Ahlfors and Beurling [4] were the first to study analytic capacity from a systematic viewpoint. Pommerenke [74] extended their results and proved some interesting new theorems; part of his work was duplicated by Ivanov [54], who also extended [55] some of the results in [4]. Vitushkin's paper [88] contains some properties of analytic capacity that relate to rational approximation. In another paper [90], Vitushkin exhibits a curious anomaly. The behavior of analytic capacity under various transformations of the domain was studied by Havin and Havinson [48]. Havinson [49] has used analytic capacity to considerable advantage in studying problems of function theory; his work is partly an extension of the work of Ahlfors and Garabedian to infinitely connected domains.

A concise treatment of the elementary properties of analytic capacity can be found in [81].

4. AC capacity. Literature on AC capacity is limited. The notion was first defined by Dolzhenko in [28]. The strongest known sufficient condition for a set to have positive AC capacity is due to Arens [7], whose result generalizes early work of Pompeiu [75], Zoretti [108], Denjoy [21],[22],[23],[24], and Urysohn [85]. It is particularly instructive to read Denjoy's papers referenced above in connection with the announcements of Pompeiu (not cited here) in the Comptes Rendus of that period; [26] contains a survey of some of these results (vol. II, pp. 631-636, 1013-1016, 1066-1067) as well as the complete text of [25] (vol. I, pp. 289-367). Besicovitch's paper [8] is also of interest here. See the book of Collingwood and Lohwater [20] for more complete references and a discussion.

5. Function algebra methods. At present, the best general references on function algebras seem to be Wermer's monograph [100], Royden's survey article [76], and Hoffman's lecture

notes [53]. [100] and [53] contain applications to rational approximation. All three references have extensive bibliographies to which the interested reader can refer. In the comments below we shall assume the language of function algebras.

Bishop [9],[14] was the first to apply the methods of functional analysis to problems of rational approximation; he gave a proof of Mergelyan's polynomial approximation theorem based on linear functionals. Glicksberg and Wermer [38] removed the remaining function theory from Bishop's argument to obtain an honest "abstract" proof: the only fact from complex variables that is needed is a result of Lebesgue [61] and Walsh [98]. A self-contained exposition of the work of Glicksberg and Wermer (plus much more) is in [103]. Carleson's synthesis [16] is probably the best available proof of Mergelyan's theorem; his treatment, based on the papers mentioned above, is abstract in spirit yet avoids the machinery of Dirichlet algebras. Glicksberg [36], using the techniques of [38], obtained an abstract proof of Mergelyan's theorem concerning rational approximation on sets of finite connectivity; using different methods Ahern and Sarason [1] obtained another proof of this result. Actually, as Garnett observed [34], Mergelyan's rational approximation theorem is a simple consequence of the theorem on polynomial approximation.

The basic paper of Bishop on peak points and the minimal boundary is [13]. Gonchar's "α-β" criterion is in [39].

Wilken generalized 9.7 in [107]; he proved that the nontrivial Gleason parts of R(X) have positive (Lebesgue planar) measure. It follows that R(X) = C(X) if and only if every point of X is a part of R(X) . See also [112].

In [62], McKissick constructed a compact set X for which R(X) is normal and yet $R(X) \neq C(X)$. Steen [82] disproved the conjecture that R(X) is antisymmetric if $R(X) \neq C(X)$ by constructing a "swiss cheese" X for which R(X) contains nonconstant real functions and yet $R(X) \neq C(X)$.

Glicksberg has proved [37] that R(X) = A(X) if and only if the real annihilating measures of these algebras coincide; further, in [35] it is shown that R(X) = A(X) if they have the same representing measures. The work of Valskii [87] is also noteworthy; it marks the first adoption by a Russian mathematician of function algebra methods for studying R(X) . Browder [15] and Wermer [104] have studied point derivations on R(X) .

Other papers worthy of note include Fisher's work [29], [30], the seminal paper of Gamelin and Rossi [32], and Wilken's paper [106]. Also, we should not fail to mention the two elegant notes of Wermer [101],[102].

6. Riemann surfaces. Sakakihara [79] seems to have been the first to consider (nontrivial) approximation on a Riemann surface; he obtained a generalization of Walsh's theorem [96].

Bishop also studied approximation on Riemann surfaces [10]. Gusman [42],[43] generalized Mergelyan's theorems on polynomial and rational approximation. Kodama [59] proceeded independently of the work of Gusman and obtained many of the same results, including the Mergelyan theorems; she also proved a "localization" theorem for surfaces, generalizing a result of Bishop's more-or-less implicit in [10]. See also [111].

7. Miscellany. There are many papers in the bibliography that do not fit easily into any of the above classifications. We shall mention some of these below.

Bishop [11],[12] studied the problem of approximating simultaneously a finite number of continuous functions by a polynomial and its derivatives. Chatskaia [17] considered the corresponding problem for rational functions.

The problem of representing a bounded analytic function as the Cauchy transform of a measure (cf. 3.13) has been studied by several authors. We mention Havin [45], Havinson [51], and Valskii [86]. Havin's papers [46],[47] also touch on this question.

The problem of approximating a continuous function on a compact set of analytic capacity zero by rational functions having a special form has also attracted attention. Such work has been done by Havin [46], Havinson [50], and Chatskaia [18],[19]. See Havinson [52] for a survey.

Bibliography

1. P. Ahern and D. Sarason, "On some hypodirichlet algebras of analytic functions", Amer. J. Math., to appear.

2. L. Ahlfors, "Bounded analytic functions", Duke Math. J. 14 (1947), 1-11. MR 9-24.

3. L. Ahlfors, "Open Riemann surfaces and extremal problems on compact subregions", Comment. Math. Helv. 24 (1950), 100-134. MR 12-90, 13-1138.

4. L. Ahlfors and A. Beurling, "Conformal invariants and function theoretic null sets", Acta Math. 83 (1950), 101-129. MR 12-171.

5. P. Appell, "Sur les fonctions uniformes d'un point analytique (X,Y)", Acta Math. 1 (1883), 109-131.

6. P. Appell, "Développements en série dans un aire limitée par des arcs de cercle", Acta Math. 1 (1883), 145-152.

7. R. Arens, "The maximal ideals of certain function algebras", Pac. J. Math. 8 (1958), 641-648. MR 22 #8315.

8. A. Besicovitch, "On sufficient conditions for a function to be analytic and on behavior of analytic functions in the neighborhood of non-isolated singular points", Proc. London Math. Soc. (2) 32 (1931), 1-9.

9. E. Bishop, "The structure of certain measures", Duke Math. J. 25 (1958), 283-289. MR 20 #5880.

10. E. Bishop, "Subalgebras of functions on a Riemann surface", Pac. J. Math. 8 (1958), 29-50. MR 20 #3300.

11. E. Bishop, "Approximation by a polynomial and its derivative on certain closed sets", Proc. Amer. Math. Soc. 9 (1958), 946-953. MR 22 #112.

12. E. Bishop, "Simultaneous approximation by a polynomial and its derivatives", Proc. Amer. Math. Soc. 10 (1959), 741-743. MR 22 #113.

13. E. Bishop, "A minimal boundary for function algebras", Pac. J. Math. 9 (1959), 629-642. MR 22 #191.

14. E. Bishop, "Boundary measures of analytic differentials", Duke Math. J. 27 (1960), 331-340. MR 22 #9621.

15. A. Browder, "Point derivations on function algebras", J. Funct. Anal. 1 (1967), 22-27.

16. L. Carleson, "Mergelyan's theorem on uniform polynomial approximation", Math. Scand. 15 (1965), 167-175. MR 33 #6368.

17. E. Sh. Chatskaia (Čackaja), "The simultaneous approximation of continuous functions by rational functions and their derivatives on some closed sets of the complex plane", Izv. Akad. Nauk. Arm. SSR ser. mat. phys. 17 (1964) 9-22. (Russian). MR 30 #389.

18. E. Sh. Chatskaia, "Some approximation problems on sets in the complex plane", Soviet Math. Dokl.6 (1965), 630-632.

19. E. Sh. Chatskaia, "On certain measures", Litovsk Mat. Sb. 5 (1965), 517-524. (Russian). MR 33 #4541.

20. E. F. Collingwood and A. J. Lohwater, The Theory of Cluster Sets, Cambridge University Press, 1966.

21. A. Denjoy, "Sur les fonctions analytiques uniformes qui restent continues sur ensemble parfait discontinu de singularités", C. R. Acad. Sci. Paris 148 (1909), 1154-1156.

22. A. Denjoy, "Sur les fonctions analytiques uniformes à singularités discontinues", C. R. Acad. Sci. Paris 149 (1909), 258-260.

23. A. Denjoy, "Sur les singularités discontinues des fonctions uniformes", C. R. Acad. Sci. Paris 149 (1909), 386-388.

24. A. Denjoy, "Sur les fonctions analytiques uniformes à singularités discontinues non isolées, C. R. Acad. Sci. Paris. 150 (1910), 32.

25. A. Denjoy, "Sur la continuité des fonctions analytiques singulières", Bull. Soc. Math. France 60 (1932), 27-105.

26. A. Denjoy, Articles et Mémoires I, II, Gauthier-Villars, Paris, 1955.

27. E. P. Dolzhenko, "Construction on a nowhere dense continuum of a nowhere differentiable function which can be expanded into a series of rational functions", Dokl. Akad. Nauk SSSR 125 (1959), 970-973. (Russian). MR 21 # 3576.

28. E. P. Dolzhenko, "Approximation on closed regions and zero sets", Soviet Math. Dokl. 3 (1962), 472-475. MR 24 #A2670.

29. S. Fisher, "Exposed points in spaces of bounded analytic functions", to appear.

30. S. Fisher, "Norm-compact sets of representing measures", to appear.

31. T. Gamelin, Algebras in the Plane, multilithed notes.

32. T. Gamelin and H. Rossi, "Jensen measures and algebras of analytic functions", in Function Algebras, ed. F. Birtel, Scott, Foresman and Co., 1966, 15-35.

33. P. Garabedian, "Schwarz's lemma and the Szëgo kernel function", Trans. Amer. Math. Soc. 67 (1949), 1-35, MR 11- 340.

34. J. Garnett, "On a theorem of Mergelyan", to appear.

35. J. Garnett and I. Glicksberg, "Algebras with the same multiplicative measures", to appear.

36. I. Glicksberg, "Dominant representing measures and rational approximation", to appear.

37. I. Glicksberg, "The abstract F. and M. Riesz theorem", J. Funct. Anal. 1 (1967), 109-122.

38. I. Glicksberg, and J. Wermer, "Measures orthogonal to a Dirichlet algebra", Duke Math. J. 30 (1963), 661-666 and 31 (1964), 717. MR 27 #6150, 29 #5124.

39. A. A. Gonchar, "On the minimal boundary of A(E)", Izv. Akad. Nauk SSSR ser. mat. 27 (1963) 949-955. (Russian). MR 27 #3812.

40. A. A. Gonchar, "Examples of non-uniqueness of analytic functions", Vestnik Moskov. Univ. ser. I mat. meh. 1 (1964), 37-43. (Russian). MR 28 #4118.

41. A. A. Gonchar and S. N. Mergelyan, "Uniform approximation by analytic and harmonic functions," in Contemporary Problems in the Theory of Analytic Functions, International Conference on the Theory of Analytic Functions, Erevan, 1965, 94-101. (Russian).

42. S. Ya. Gusman, "Uniform approximation of continuous functions on Riemann surfaces", Soviet Math. Dokl. 1 (1960), 105-107. MR 22 #11124.

43. S. Ya. Gusman, "The uniform approximation of continuous functions on Riemann surfaces I, II", Izv. Vyssh. Uchebn. Zaved. Mat. 18 (1960), 43-51 and 20 (1961), 44-54. (Russian). MR 24 #A2038a,b.

44. F. Hartogs and A. Rosenthal, "Über Folgen analytischen Funktionen", Math. Ann. 104 (1931) 606-610.

45. V. P. Havin, "On analytic functions representable by an integral of Cauchy-Stieltjes type", Vestnik Leningrad. Univ. ser. mat. meh. astron. 13 (1958), 66-79. (Russian). MR 20 #1762.

46. V. P. Havin, "On the space of bounded regular functions", Soviet Math. Dokl. 1 (1960), 202-204. MR 22 #11277.

47. V. P. Havin, "On the space of bounded regular functions", Sibirsk. Mat. Zh. 2 (1961), 622-638. (Russian). MR 25 #2425.

48. V. P. Havin and S. Ya. Havinson, "Some estimates of analytic capacity", Soviet Math. Dokl. 2 (1961), 731-734. MR 24 #1379.

49. S. Ya. Havinson, "Analytic capacity of sets, joint nontriviality of various classes of analytic functions and the Schwarz lemma in arbitrary domains", Amer. Math. Soc. Translations ser. 2 vol. 43, 215-266. MR 25 #182.

50. S. Ya. Havinson, "Approximation on sets of zero analytic capacity", Soviet Math. Dokl. 1 (1960), 205-207. MR 23 #A1045.

51. S. Ya. Havinson, "Some remarks on integrals of Cauchy-Stieltjes type", Litovsk Mat. Sb. 2 (1963) 281-288. (Russian). MR 28 #1280.

52. S. Ya. Havinson, "On the representation and approximation of functions on thin sets", in Contemporary Problems in the Theory of Analytic Functions, International Conference on the Theory of Analytic Functions, Erevan, 1965, 314-318. (Russian).

53. K. Hoffman, "Lectures on sup norm algebras", in Summer School on Topological Algebra Theory, Bruges, 1966. (mimeographed notes).

54. L. D. Ivanov, "On the analytic capacity of linear sets", Uspehi Mat. Nauk 17 (108) (1962), 143-144. (Russian). MR 29 #264.

55. L. D. Ivanov, "On Denjoy's conjecture", Uspehi Mat. Nauk 18 (112) (1963), 147-149. (Russian). MR 28 #236.

56. M. V. Keldysh, "Sur l'approximation des fonctions analytiques dans des domaines fermés, Mat. Sb. N. S. 8 (50) (1940), 137-148. MR 2-188.

57. M. V. Keldysh, "On the solvability and stability of the Dirichlet problem", Uspehi Mat. Nauk 8 (1941), 171-231. (Russian) MR 3-123.

58. M. V. Keldysh, "Sur la representation par des séries de polynomes des fonctions d'une variable complexe dans de domaines fermés", Mat. Sb. N. S. 16 (58) (1945), 249-258. MR 7-285.

59. L. Kodama, "Boundary measures of analytic differentials and uniform approximation on a Riemann surface", Pac. J. Math. 15 (1965), 1261-1277. MR 32 #7740.

60. M. A. Lavrentieff, Sur les fonctions d'une variable complexe représentables par des séries de polynomes, Hermann, Paris, 1936.

61. H. Lebesgue, "Sur le problem de Dirichlet", Rend. Circ. Mat. di Palermo 29 (1907), 371-402.

62. R. McKissick, "A non-trivial sup norm algebra", Bull. Amer. Math. Soc. 69 (1963), 391-395. MR 26 #4166

63. M. S. Melnikov, "Analytic capacity and the Cauchy integral", Soviet Math. Dokl. 8 (1967), 20-23.

64. M. S. Melnikov, "A bound for the Cauchy integral along an analytic curve", Mat. Sb. 71 (113) (1966), 503-515. (Russian).

65. S. N. Mergelyan, "On a theorem of M. A. Lavrent'ev", in Series and Approximation (Amer. Math. Soc. Translations) ser. I vol. 3, 281-286. MR 12-814, 14-858.

66. S. N. Mergelyan, "On the representation of functions by series of polynomials on closed sets", ibid., 287-293. MR 13-23, 14-858.

67. S. N. Mergelyan, "Uniform approximation to functions of a complex variable", ibid., 294-391. MR 14-547, 15-612.

68. S. N. Mergelyan, "On some results in the theory of uniform and best approximation by polynomials and rational functions", foreword to Russian edition of [99].

69. S. N. Mergelyan, "Certain classes of sets and their applications", Soviet Math. Dokl. 2 (1961), 590-593. MR 24 #A1408.

70. Z. Nehari, "Bounded analytic functions", Bull. Amer. Math. Soc. 57 (1951), 354-366. MR 13-222.

71. P. Painlevé, "Sur les lignes singulières des fonctions analytiques", Ann. Fac. Sci. Univ. Toulouse 2 (1888).

72. P. Painlevé, "Observations au sujet de la Communication précédent", C. R. Acad. Sci. Paris 148 (1909), 1156-57.

73. R. Phelps, Lectures on Choquet's Theorem, D. Van Nostrand Company, Inc., Princeton, N. J., 1966.

74. Ch. Pommerenke, Über die analytische Kapazität", Archiv der Math. 11 (1960), 270-277. MR 22 #5740c.

75. D. Pompeiu, "Sur la continuité des fonctions de variables complexes", Ann. Fac. Sci. Univ. Toulouse (2) 7 (1905), 264-315.

76. H. Royden, "Function algebras", Bull. Amer. Math. Soc. 69 (1963), 281-298. MR 26 #6817.

77. H. Royden, "The boundary values of analytic and harmonic functions", Math Zeitschr. 78 (1962), 1-24. MR 25 #2190.

78. C. Runge, "Zur Theorie der eindeutigen analytischen Funktionen", Acta Math 6 (1885), 228-244.

79. K. Sakakihara, "Meromorphic approximations on Riemann surfaces", J. Inst. Polytech. Osaka City Univ. ser. A math. 5 (1954), 63-70. MR 16-687.

80. M. Schiffer, "The span of multiply connected domains", Duke Math J. 10 (1943), 209-216. MR 4-271.

81. V. I. Smirnov and N. A. Lebedev, The Constructive Theory of Functions of a Complex Variable, Nauka, Leningrad, 1964. (Russian). MR 30 #2152.

82. L. Steen, "On uniform approximation by rational functions", Proc. Amer. Math. Soc. 17 (1966), 1007-1011. MR 33 #7561.

83. V. A. Tonyan, "On approximation of continuous function on sets separating the plane", Dokl. Akad. Nauk Arm. SSR 12 (1950), 33-36. (Russian). MR 14-548.

84. M. Tsuji, Potential Theory in Modern Function Theory, Maruzen Co., Ltd., Tokyo, 1959. MR 22 #5712.

85. P. Urysohn, "Sur une fonction analytique partout continue", Fund. Math. 4 (1923), 144-150.

86. R. E. Valskii, "Remarks on bounded functions representable by an integral of the Cauchy-Stieltjes type", Siberian Math. J. 7 (1967), 202-209. MR 33 #4829.

87. R. E. Valskii, "Gleason parts for algebras of analytic functions and measures orthogonal to these algebras", Soviet Math. Dokl. 8 (1967), 300-303.

88. A. G. Vitushkin, "Analytic capacity of sets and some of its properties", Dokl. Akad. Nauk SSSR 123 (1958), 778-781. (Russian). MR 21 #2056.

89. A. G. Vitushkin, "Some theorems on the possibility of uniform approximation of continuous functions by analytic functions", Dokl. Akad. Nauk. SSSR 123 (1958), 958-962. (Russian). MR 21 #2057.

90. A. G. Vitushkin, "Example of a set of positive length but of zero analytic capacity", Dokl. Akad. Nauk SSSR 127 (1959), 246-249. (Russian). MR 22 #9607.

91. A. G. Vitushkin, "Necessary and sufficient conditions a set should satisfy in order that any function continuous on it can be approximated uniformly by analytic or rational functions", Dokl. Akad. Nauk. SSSR 128 (1959), 17-20. (Russian) MR 22 #775.

92. A. G. Vitushkin, "On a problem of Denjoy", Izv. Akad. Nauk SSSR ser. mat. 28 (1964), 745-756. (Russian). MR 29 #6024.

93. A. G. Vitushkin, "Approximation of function by rational functions", Soviet Math. Dokl. 7 (1966), 1582-1585.

94. A. G. Vitushkin, "Necessary and sufficient conditions on a set in order that any continuous function analytic at the interior points of the set may admit of uniform approximation by rational functions", Soviet Math. Dokl. 7 (1966), 1622-1625

95. A. G. Vitushkin, "A bound for the Cauchy integral", Mat. Sb. 71 (113) (1966), 515-535.

96. J. Walsh, "Über die Entwicklung einer analytischen Funktion nach Polynomen", Math. Ann. 96 (1926), 430-436.

97. J. Walsh, "Über die Entwicklung einer Funktion einer komplexen Veränderlichen nach Polynomen", Math. Ann. 96 (1926), 437-450.

98. J. Walsh, "The approximation of harmonic functions by harmonic polynomials and harmonic rational functions", Bull. Amer. Math. Soc. 35 (1929), 499-544.

99. J. Walsh, Interpolation and Approximation by Rational Functions in the Complex Domain, Colloq. Publ. Amer. Math. Soc. vol. 20, New York, 1956.

100. J. Wermer, "Banach algebras and analytic functions", Advances in Math. 1 (1961) fasc. 1, 51-102. MR 26 #629.

101. J. Wermer, "Approximation on a disc", Math. Ann. 155 (1964), 331-333. MR 29 #2670.

102. J. Wermer, "Polynomially convex discs", Math. Ann. 158 (1965), 6-10. MR 30 #5158.

103. J. Wermer, Seminar über Funktionen-Algebren, Lecture notes in math., Springer Verlag, Berlin, 1964.

104. J. Wermer, "Bounded point derivations on certain Banach algebras", J. Funct. Anal. 1 (1967), 28-36.

105. N. Wiener, "The Dirichlet problem", J. Math. Phys. 3 (1924), 127-146.

106. D. Wilken, "Approximate normality and function algebras on the interval and the circle", in Function Algebras, ed. F. Birtel, Scott, Foresman and Co., 1966, 98-111. MR 33 #4712.

107. D. Wilken, Lebesgue measure for parts of R(X)", Proc. Amer. Math. Soc. 18 (1967), 508-512.

108. L. Zoretti, "Sur les fonctions analytiques uniformes qui possèdent un ensemble parfait discontinu de points singuliers", J. Math. Pures Appl. (6) 1 (1905), 1-51.

Bibliographical Addenda

109. L. Carleson, "On null-sets for continuous analytic functions", Arkiv för Mat. 1 (1950), 311-318. MR 13-23.

110. E. P. Dolzhenko, "The removability of singularities of analytic functions", Uspehi Mat. Nauk 18 (1963) (112), 135-142. (Russian). MR 27 #5898.

111. T. Gamelin and G. Lumer, "The universal Hardy class", to appear.

112. M. S. Melnikov, "Structure of the Gleason part of the algebra R(E)", Funct. Anal. Appl. 1 (1967), 84-86.

113. D. Wilken, "The support of representing measures for R(X)", to appear.

114. P. C. Curtis, "Peak points for algebras of analytic functions", to appear.

Lecture Notes in Mathematics

Bisher erschienen/Already published

Vol. 1: J. Wermer, Seminar über Funktionen-Algebren. IV, 30 Seiten. 1964. DM 3,80 / $ 0.95

Vol. 2: A. Borel, Cohomologie des espaces localement compacts d'après J. Leray. IV, 93 pages. 1964. DM 9,– / $ 2.25

Vol. 3: J. F. Adams, Stable Homotopy Theory. 2nd. revised edition. IV, 78 pages. 1966. DM 7,80 / $ 1.95

Vol. 4: M. Arkowitz and C. R. Curjel, Groups of Homotopy Classes. 2nd. revised edition. IV, 36 pages. 1967. DM 4,80 / $ 1.20

Vol. 5: J.-P. Serre, Cohomologie Galoisienne. Troisième édition. VIII, 214 pages. 1965. DM 18,– / $ 4.50

Vol. 6: H. Hermes, Eine Termlogik mit Auswahloperator. IV, 42 Seiten. 1965. DM 5,80 / $ 1.45

Vol. 7: Ph. Tondeur, Introduction to Lie Groups and Transformation Groups. VIII, 176 pages. 1965. DM 13,50 / $ 3.40

Vol. 8: G. Fichera, Linear Elliptic Differential Systems and Eigenvalue Problems. IV, 176 pages. 1965. DM 13.50 / $ 3.40

Vol. 9: P. L. Ivănescu, Pseudo-Boolean Programming and Applications. IV, 50 pages. 1965. DM 4,80 / $ 1.20

Vol. 10: H. Lüneburg, Die Suzukigruppen und ihre Geometrien. VI, 111 Seiten. 1965. DM 8,– / $ 2.00

Vol. 11: J.-P. Serre, Algèbre Locale. Multiplicités. Rédigé par P. Gabriel. Seconde édition. VIII, 192 pages. 1965. DM 12,– / $ 3.00

Vol. 12: A. Dold, Halbexakte Homotopiefunktoren. II, 157 Seiten. 1966. DM 12,– / $ 3.00

Vol. 13: E. Thomas, Seminar on Fiber Spaces. IV, 45 pages. 1966. DM 4,80 / $ 1.20

Vol. 14: H. Werner, Vorlesung über Approximationstheorie. IV, 184 Seiten und 12 Seiten Anhang. 1966. DM 14,– / $ 3.50

Vol. 15: F. Oort, Commutative Group Schemes. VI, 133 pages. 1966. DM 9,80 / $ 2.45

Vol. 16: J. Pfanzagl and W. Pierlo, Compact Systems of Sets. IV, 48 pages. 1966. DM 5,80 / $ 1.45

Vol. 17: C. Müller, Spherical Harmonics. IV, 46 pages. 1966. DM 5,– / $ 1.25

Vol. 18: H.-B. Brinkmann und D. Puppe, Kategorien und Funktoren. XII, 107 Seiten. 1966. DM 8,– / $ 2.00

Vol. 19: G. Stolzenberg, Volumes, Limits and Extensions of Analytic Varieties. IV, 45 pages. 1966. DM 5,40 / $ 1.35

Vol. 20: R. Hartshorne, Residues and Duality. VIII, 423 pages. 1966. DM 20,– / $ 5.00

Vol. 21: Seminar on Complex Multiplication. By A. Borel, S. Chowla, C. S. Herz, K. Iwasawa, J.-P. Serre. IV, 102 pages. 1966. DM 8,– / $ 2.00

Vol. 22: H. Bauer, Harmonische Räume und ihre Potentialtheorie. IV, 175 Seiten. 1966. DM 14,– / $ 3.50

Vol. 23: P. L. Ivănescu and S. Rudeanu, Pseudo-Boolean Methods for Bivalent Programming. 120 pages. 1966. DM 10,– / $ 2.50

Vol. 24: J. Lambek, Completions of Categories. IV, 69 pages. 1966. DM 6,80 / $ 1.70

Vol. 25: R. Narasimhan, Introduction to the Theory of Analytic Spaces. IV, 143 pages. 1966. DM 10,– / $ 2.50

Vol. 26: P.-A. Meyer, Processus de Markov. IV, 190 pages. 1967. DM 15,– / $ 3.75

Vol. 27: H. P. Künzi und S. T. Tan, Lineare Optimierung großer Systeme. VI, 121 Seiten. 1966. DM 12,– / $ 3.00

Vol. 28: P. E. Conner and E. E. Floyd, The Relation of Cobordism to K-Theories. VIII, 112 pages. 1966. DM 9.80 / $ 2.45

Vol. 29: K. Chandrasekharan, Einführung in die Analytische Zahlentheorie. VI, 199 Seiten. 1966. DM 16.80 / $ 4.20

Vol. 30: A. Frölicher and W. Bucher, Calculus in Vector Spaces without Norm. X, 146 pages. 1966. DM 12,– / $ 3.00

Bitte wenden / Continued

Vol. 31: Symposium on Probability Methods in Analysis. Chairman: D.A.Kappos. IV, 329 pages. 1967. DM 20,– / $ 5.00

Vol. 32: M. André, Méthode Simpliciale en Algèbre Homologique et Algèbre Commutative. IV, 122 pages. 1967. DM 12,– / $ 3.00

Vol. 33: G. I. Targonski, Seminar on Functional Operators and Equations. IV, 110 pages. 1967. DM 10,– / $ 2.50

Vol. 34: G. E. Bredon, Equivariant Cohomology Theories. VI, 64 pages. 1967. DM 6,80 / $ 1.70

Vol. 35: N. P. Bhatia and G. P. Szegö, Dynamical Systems: Stability Theory and Applications. VI, 416 pages. 1967. DM 24,– / $ 6.00

Vol. 36: A. Borel, Topics in the Homology Theory of Fibre Bundles. VI, 95 pages. 1967. DM 9,– / $ 2.25

Vol. 37: R. B. Jensen, Modelle der Mengenlehre. X, 176 Seiten. 1967. DM 14,– / $ 3.50

Vol. 38: R. Berger, R. Kiehl, E. Kunz und H.-J. Nastold, Differentialrechnung in der analytischen Geometrie. IV, 134 Seiten. 1967. DM 12,– / $ 3.00

Vol. 39: Séminaire de Probabilités I. II, 189 pages. 1967. DM 14,– / $ 3.50

Vol. 40: J. Tits, Tabellen zu den einfachen Lie Gruppen und ihren Darstellungen. VI, 53 Seiten. 1967. DM 6,80 / $ 1.70

Vol. 41: R. Hartshorne, Local Cohomology. VI, 106 pages. 1967. DM 10,– / $ 2.50

Vol. 42: J. F. Berglund and K. H. Hofmann, Compact Semitopological Semigroups and Weakly Almost Periodic Functions. VI, 160 pages. 1967. DM 12,– / $ 3.00

Vol. 43: D. G. Quillen, Homotopical Algebra. VI, 157 pages. 1967. DM 14,– / $ 3.50

Vol. 44: K. Urbanik, Lectures on Prediction Theory. IV, 50 pages. 1967. DM 5,80 / $ 1.45

Vol. 45: A. Wilansky, Topics in Functional Analysis. VI, 102 pages. 1967. DM 9,60 / $ 2.40

Vol. 46: P. E. Conner, Seminar on Periodic Maps. IV, 116 pages. 1967. DM 10,60 / $ 2.65

Vol. 47: Reports of the Midwest Category Seminar. IV, 181 pages. 1967. DM 14,80 / $ 3.70

Vol. 48: G. de Rham, S. Maumary and M. A. Kervaire, Torsion et Type Simple d'Homotopie. IV, 101 pages. 1967. DM 9,60 / $ 2.40

Vol. 49: C. Faith, Lectures on Injective Modules and Quotient Rings. XVI, 140 pages. 1967. DM 12,80 / $ 3.20